PRAISE FOR *FRIENDS, FOLLOWERS AND THE FUTURE*

"This is a timely book about a vital subject: How do we get information and is it reliable? With a 'cold eye,' author Rory O'Connor shows how traditional journalism cheapened its value by sabotaging its trust, and how the digital revolution wonderfully democratizes information yet often removes the journalistic curator, creating more noise, more ME and less WE news. If you want to understand the future of news, its opportunities and its pitfalls, read this book."—Ken Auletta, author and *New Yorker* media writer

"Anyone who cares about the impact of the digital information revolution on democracy and culture can't afford to miss *Friends, Followers and the Future*—a story that moves as swiftly as the dizzying pace of change itself. Rory O'Connor combines journalistic integrity with a passionate belief in the power of ordinary people to change the world. Depending on your stake in the outcome, you will find this book inspiring, scary, or perhaps a bit of both."—Andrew Heyward, former President of CBS News

"With laser-like accuracy Rory O'Connor spotlights the key challenges and opportunities in the world of news and information, where technology has upended the old rules of how media is created and consumed. O'Connor's wise, savvy *Friends, Followers and the Future* is an essential examination of how social media is transforming the lives of individuals and society at large. Read it and share it."—J. Max Robins, Vice President/Executive Director, The Paley Center for Media

FRIENDS, FOLLOWERS
and the
FUTURE

How Social Media are Changing Politics,
Threatening Big Brands, and Killing
Traditional Media

Rory O'Connor

City Lights Books • San Francisco

To Ciaran and Aidan, two Digital Natives

The brand that increasingly matters is the one called 'my friend.'
> —Mark Lukasiewicz, NBC News executive

More and more we will be looking at our Facebook feed to see what friends have posted. That will be how we queue up what is important and credible.
> —BJ Fogg, Stanford University professor

He not busy being born is busy dying.
> —Bob Dylan

eISBN 978-0-87286-562-4 (ebook)

Library of Congress Cataloging-in-Publication Data
O'Connor, Rory, 1951-
 Friends, followers, and the future : how social media are changing
politics, threatening big brands, and killing traditional media / Rory
O'Connor.
    p. cm.
 ISBN 978-0-87286-556-3 (paper)
1. Social media. 2. Social media. 3. Social media—Political aspects.
4. Social media—Economic aspects. I. Title.
 HM742.O26 2012
 302.3—dc23

                    2012005506

City Lights Books are published at the City Lights Bookstore
261 Columbus Avenue, San Francisco, CA 94133
www.citylights.com

Contents

Introduction: Word of Mouse

New York, November 2011

As I write, demonstrators all over the world have taken to the streets to protest against social, economic and political systems—many widely regarded as democratic—that simply aren't serving their needs. From Cairo to Athens, from New Delhi to Tel Aviv, and from Madrid to a just few miles away here in Manhattan, indignant citizens are rising up and raising their voices to send loud messages of dissatisfaction to the powers-that-be: Big Government, Big Business, and Big Media alike.

Growing networks of "ordinary people," many of them young and feeling deeply disenfranchised, are beginning to exert extraordinary influence on societies worldwide. They are using emerging media forms to bypass state censorship, outpace traditional news organizations, and compel corporations and governments alike to listen to and act on their demands. Deeply disruptive new media tools, which now enable them to produce and distribute news and information widely, inexpensively, and efficiently, also help them to attract large audiences, to affect the rise and fall of policies, politicians and entire governments and even to revolution-ize entire industries.

Here in the United States—long regarded as the global bastion of capitalist democracy—an ongoing "occupation" of Wall Street and the financial district at the epicenter of the world's continuing economic crisis, continues to grow in impact and importance. Initially ignored or ridiculed by the powerful global media corporations also headquartered here, the demonstrators responded by employing their own media—including viral emails, blogs, social networks like Facebook and other young social platforms, such as the video sharing site YouTube and the micro-blogging Twitter service—to spread their message by word of mouse.

Dismayed by the corporate media's spotty, cynical cov-erage and upset at what they viewed as a lack of authentic information being made available about their movement, organizers relied instead on telling their own story, through diverse media that included an online livestream of the oc-cupation, websites, a newspaper, and much, much more. Some in the tech-savvy crowd posted commentary, photo-graphs and video to Twitter and YouTube, even when under arrest. Their friends and followers quickly redistributed the pictures, posts and tweets emanating from what soon be-came renowned as "Occupy Wall Street." Within days the more established media were forced to change their conde-scending stance and begin to cover the protests in a more comprehensive and respectful manner.

Similar protests soon sprouted elsewhere. The loose-knit campaign that began in New York spread to dozens of other cities across the country, "with protesters camped out in Los Angeles near City Hall, assembled before the Federal Reserve Bank in Chicago and marching through downtown Boston to rally against corporate greed, unemployment and the role of financial institutions in the economic crisis," as the Agence France-Presse wire service reported. "With little organization and a reliance on Facebook, Twitter and Google groups to share methods, the Occupy Wall Street campaign, as the prototype in New York is called, has clearly tapped into a deep vein of anger, experts in social movements said, bringing longtime crusaders against globalization and professional anarchists together with younger people frustrated by poor job prospects."

When New York police arrested more than 700 people on the Brooklyn Bridge on October 1, 2011, it only galvanized their campaign and provided momentum for new rallies and other encampments in cities as disparate as Baltimore, Memphis, Minneapolis, and even the nation's capital, where an Occupy DC protest began in a park near the White House. While tens of thousands of union members and other progressives joined the demonstrators to denounce the power, wealth, and indifference of America's major financial institutions, Occupy Together, an unofficial hub for the protests, listed sites for hundreds of future demonstrations, including some in Europe and Japan.

Nicholas Kulish chronicled the growing demonstrations in a *New York Times* report published ten days after the occupation of Wall Street began, noting that the protesters' complaints "range from corruption to lack of affordable housing and joblessness, common grievances the world over. But from South Asia to the heartland of Europe and now even to Wall Street, these protesters share something else: wariness, even contempt, toward traditional politicians and the democratic political process they preside over."

Why are people all over the world coming to similar conclusions and taking to the streets all at once to express them? The answer should be obvious: *they feel their social systems have abandoned them.* As a result, citizens of all ages, but especially young people, are throwing off old-style, top-down constructs like political parties, traditional media and corporate brands to adopt instead the more participatory and far less hierarchical ways of the Web.

Thus the protest movements in democracies such as the United States and Israel have much in common with those that toppled authoritarian governments during the Arab Spring of 2011. In each case, a disenchanted populace moved to create its own online space, using tools such as wikis, social networks, Twitter and the like to form instant networks of like-minded people.

"You're looking at a generation of 20- and 30-year-olds who are used to self-organizing," Yochai Benkler, a director of the Berkman Center for Internet and Society at Harvard University, explained to the *New York Times.* "They believe life can be more participatory, more decentralized, less dependent on the traditional models of organization, either in the state or the big company. Those were the dominant ways of doing things in the industrial economy, and they aren't anymore."

After the collapse of the Soviet Union and the end of the Cold War, which had followed closely on the heels of the Second World War's hotter antagonisms, triumphant cheerleaders for the democratic capitalist system celebrated victory over authoritarian communism by famously declaring an "end to history." They pronounced with unexamined certitude the fact that capitalistic economics and globalization, combined with the Western world's democratic institutions, represented the only way forward to the future. But repeated economic collapses—1997's Asian financial "flu," the popping of the "Internet bubble" a few years later, the "Great Recession" of 2007–8, and the continuing debt crisis still roiling global markets—soon put their theory to the test. The subsequent paralysis of policy makers and their inability (or outright refusal) to protect their citizens from the ill effects of capitalism's crises left many frustrated at best with stale political choices they saw as mere leftovers of an era that had already ended. Is it any surprise then that when offered powerful and cost-free new tools of communication, many have now moved to create their own media, brands, and modes of governance?

As powerful as they are, the protests of 2011—from the Arab Spring that toppled dictators in Tunisia, Egypt, and Libya to the American Autumn and the Occupy movement—are but the most current and specific manifestations of an evolving global phenomenon now affecting nearly every aspect of modern life. In point of fact, there really *is* a revolution going on—not just a political one, but a *digital information revolution*—that encompasses nearly every aspect of the way we live, work, play, vote, govern, and do business, and which is rapidly and radically transforming how we communicate.

Internet technology is constantly evolving new tools, from blogging to tweeting and from social networking to link sharing, that enable us all to produce and distribute diverse streams of news and information. For the first time in history, we now enjoy myriad ways through which we all can make our message, as well as that of others, much more accessible and amplified. The impact of this information revolution goes far beyond just the political realm, even as it has led to such previously unimaginable events as, for example, the shockingly rapid fall of Egypt's longtime and once seemingly all-powerful leader Hosni Mubarak.

The accelerating impact of the Internet, and the powerful social media that have begun to dominate its use, are now rapidly transforming our political, commercial and communications environments, while profoundly affecting the future of governance and the very nature of democracy itself. No longer passive consumers of profit-driven centralized news, government edicts, corporate advertising, and other top-down forms of information products, active users of online social media increasingly produce and break news themselves from the bottom up—while also becoming trusted curators of and commentators on events that direct-ly affect our lives. Although this ongoing revolution may not be televised by the legacy broadcast media, it is certainly being *digitized*, and then distributed and redistributed at the speed of light to every corner of our increasingly connected world. Considered together, the newly emerging commu-nications technologies signal a huge shift in how we now find, consume and interact with news and information of all types.

Share and share alike—the sheer numbers say it all, as literally billions of people flock to social media sites of all sorts to communicate what they deem important to their lives and that of people within their new online networks, known in social parlance as "friends" and "followers." The ramifications of this seismic shift are tremendous. Not long ago, highly centralized sources—the so-called media brands—were almost universally trusted and thought by many to be virtually omniscient. (Those of a certain age will recall *CBS News* anchorman Walter Cronkite's nightly sign-off, "And that's the way it is!") As we now move to a new era of decentralized, distributed and networked social media, however, such old realities evanesce.

In this dawning age of media abundance, access to the means of production and distribution is no longer limited to a self-ordained priesthood of professionals. Instead, as networked technologies proliferate, new methods of creat-ing content and new channels to distribute it have become available *to everyone* and *between everyone*. With barriers collapsing, power no longer resides solely with the legacy media brands that are still grasping for control with their old methods and channels.

Along with losing their previously privileged position, legacy media have also lost the trust of many of their customers—largely as a result of their poor performance over time. As online outlets like Facebook, Twitter, and YouTube and others supplement (and begin to supplant) both legacy outlets and Google with its vaunted search engine as distributors of credible news and information, they also alter our relationships with older media. With our longstanding dependence on them diminished, the legacy media and their corporate advertisers can no longer rely mainly on brand power to inspire trust and confidence.

As previous structures vanish and the primacy of legacy media brands fades, revenue models, and commercial relationships of the past disintegrate. Entire industries such as television, film, newspapers, magazines, (and yes, even book publishing) are being radically, sometimes even fatally, disrupted. The rise of the new social media will inevitably lead to the obsolescence and eventual death or transformation of any and all brands that fail to embrace and adapt to this quickly morphing mediascape.

What changed, when, how, and why? Over the course of the last decade, the emerging social media—most of which didn't even exist before the turn of the century—abruptly altered our use of the Internet, itself still a relatively young medium. Interactive "Web 2.0" platforms such as

Facebook, YouTube, and Twitter experienced phenomenal growth by freely offering useful new tools and technologies. These platforms quickly came to dominate the amount of time users spent online, taking the place of such previously popular Internet pastimes as email and search. In a breathtakingly short period, social media became touchstones of modern communications and culture—and in the process upended entire industries, changed cultural norms, and disrupted both national and then international politics and, to a lesser degree, governance. As social media made billionaires of young entrepreneurs like Facebook's co-founder Mark Zuckerberg, while inspiring such best selling books as *The Facebook Effect* and Hollywood blockbusters as *The Social Network*, they also became prominent news sources for an ever-increasing number of people who were simultaneously losing their faith or simply no longer paying attention to long-established media brands.

Almost overnight, the previous century's centralized, one-way media reality of limited channels for the distribution of news and information—and thus of limited access to it—was transformed, as the new methods of online media distribution became available. While traditional media were busy dying, new networks were being born, offspring of an emergent link economy itself born of Google's once-revolutionary search technology.

Meanwhile Internet users with common interests in such topics as music, photography and current events began to coalesce online, initially on such platforms as Friendster and MySpace, and later on YouTube, Flickr, Facebook, Twitter, and others. At the same time, a host of external financial and structural challenges beset and began to overwhelm the traditional news media, which were also racked with a variety of self-imposed problems, ranging from declining audience interest to increasing political partisanization and even complicity in ginning up unpopular and unnecessary wars.

One result was a severe erosion in public confidence in "old media" forms. By the start of the new century's second decade, it had become commonplace when searching for credible news and information to rely at least in part on friends and followers within these new online networks. At the same time, the dynasties and dinosaurs of the legacy media and traditional journalism were besieged by twin crises of diminishing resources and questionable credibility.

This exciting, messy, and chaotic digital media revolution, with its flood of unmediated news and information, unaddressed trust-and-credibility questions, emerging social platforms and their highly disruptive effect on brands, from long-established legacy firms such as *NBC News* or the *New York Times* to new media newcomers like Google, and the attendant commercial, communications, and political turmoil, is the story of our time—and the subject of this book. In less than a decade these new platforms leapt into prominence and became major global conduits of news, information, and social action. The emergence of our online social networks and their new media tools and technologies directly imperil every legacy media industry. With "social" now beginning to replace "search" as a main focus on online activity, these new online titans also pose an increasing existential threat to such other Internet behemoths as Google—the current King-of-the-Hill and long thought, as the world's leading brand, to be invulnerable and virtually immune to competition.

And as we have seen, just months after the uprising in Cairo's Tahrir Square toppled the old authoritarian order and became synonymous with radical transformation, "Facebook-driven" protests all over the globe threatened equally to upend many other societal behemoths. Amidst common and continuing complaints from consumers and citizens alike about credibility, trust, authenticity, transparency and the nature of democracy itself, the overriding issue is not specifically political or economic in nature, but ultimately one of power relations. Who will determine our future—the centralized dominating forces of the past, or we the people?

The answer echoes from unlikely places. The occupation in Egypt's capital during Arab Spring had a surprising and unanticipated impact in Israel, with citizens there citing the events in Tahrir Square as an inspiration for protests in their own capital of Tel Aviv. As Moshe Gant, a 35-year-old Israeli business analyst who came to support the demonstrators, told the *Times*, "Religious Jews like to think of us as a light unto nations, meaning that others will learn from us, but this time we have learned from the nations around us that change can come from people power." Protester El-dad Yaniv amplified Gant's remarks. "This is the first time that instead of fighting against the Arabs we are fighting for something—our life and that of our children. The old right and left are fading. This country needs a new left, its own New Deal."

Meanwhile back here in New York City, what started less than two months ago as a small demonstration, also inspired by the events in Tahrir Square, has already grown into a national movement. As the richest and most powerful 1 percent becomes ever richer, the "other 99%" is taking to the streets in response, protesting their country's economic imbalance, powerful corporations, and financial entities— and its political and media systems, which are no longer addressing their needs and thus face the risk of increasing irrelevance.

The tipping point is upon us. The unprecedented power of the emerging social media is helping people connect online and in the streets to push the entire world over the edge of change. Watch out, Big Media, Big Business, and Big Government—here come our friends, our followers, and our future!

The Rise of Social Media

In September 2008 I moved back to my old stomping grounds and adopted hometown of Cambridge, Massachusetts to begin a fellowship at Harvard University's Joan Shorenstein Center on the Press, Politics and Public Policy. The Shorenstein fellowship would enable me, for the first time in years, to step back from the daily hurly-burly of writing, filmmaking, and managing Globalvision, the small Manhattan-based independent media production firm I had co-founded two decades earlier.

My office at the Kennedy School looked out onto a radically different Harvard Square from the one I had first seen as a student in 1968. It was an ideal perch from which to research, reflect, and respond to the massive changes then roiling my chosen profession of journalism. Harvard's cachet drew a parade of powerful players to the Kennedy School from the increasingly interconnected worlds of media, politics, and technology. One result was a seemingly endless smorgasbord of speeches and seminars, interviews and panel discussions, brown-bag lunches and formal dinners.

Like many journalists of my generation, my initial motivation for working in the media had been born of the turbulence and ferment of the 1960s and a resultant if vague desire to "do something about the problems of the world." Soon, however, I came to realize that *the media was one of the problems of the world.*

It's no wonder that only one of four Americans now trusts the accuracy of the news and information he or she receives. Let's be frank, after all; we live in an age of media scams and scandals, of manipulated images and rented opinions, of "information dominance" and partisan pay-per-post, and of fake news and staged events. . . . From Comedy Central's joker-journalists on *The Daily Show* and *Colbert Report* to faux news on Fox News all the way to phony video news releases and "grassroots" public relations blogs that are actually "AstroTurf" fronts for huge corporations, and from government-and-corporate-sponsored syndicated commentary to Pentagon propaganda posing as authentic journalism to undocumented docudramas, wherever and whenever you look, examples of mainstream media make-believe abound.

The new media world of the Internet, with its emphasis on speed and immediacy, had in some respects only made this crisis of confidence worse. Opening up and democratizing creation and distribution of content also opened a Pandora's box of credibility concerns, as unfiltered, unverified and often untrue claims—such as the persistent meme that President Barack Obama is Muslim, or that a presidential visit to India cost $200 million dollars a day and involved a tenth of all the ships in the U.S. Navy—were regularly asserted, swallowed whole, regurgitated and then repeated, over and over, until accepted by many as facts.

One result has been a generalized mistrust in our ability to find credible news and information. Many factors contribute to the malaise: years of consolidation of news media by multinational media/entertainment conglomerates; subsequent budgetary and profit-making pressures leading to layoffs, buyouts, cutbacks, and shortcuts within traditional news organizations; the concomitant spread of cable television's cheap-to-produce, highly partisan and factually challenged opinions-as-news programs; rapid technological advances and the digital information revolution's disruption of legacy media's longstanding revenue models and distribution patterns; its explosion of sometimes spurious online outlets. . . . The list is already long, and growing still.

Whatever the causes, the results seem clear: in today's rapidly morphing media landscape, sloppy, gossipy and sometimes just plain manufactured alleged journalism is increasingly treated as legitimate news, opinions are presented as truths, and misinformation spreads like wildfire. It is harder than ever to separate actual opinion from outright fiction. Many consumers bemoan what they now term the "corporate media," seeing it as more interested in private profit than public interest; at the same time many professionals decry so-called "citizen journalism," fearing that it will inevitably lead to a loss of quality. Issues of accountability and ethics in gathering and disseminating news and information are now more critical than ever, and the need for mechanisms and filters that can assist us in identifying verifiable, trustworthy news and information of high quality is thus greater. The lack of trust between the media and the populace is at an all-time high.

Ironically, we live in a media-saturated era, one in which news and information from a wide range of sources is readily available to many of us for the first time in history. Although such unparalleled information access can be empowering, it is also quite disruptive and presents a unique set of issues and challenges both to journalists and to society as a whole. Central among them is that of trust. Facing an unfiltered torrent of news and information—powered by dazzling new tools and technologies and distributed at the speed of light—how can any of us be sure that the news we see and hear is true? How can we find information we can trust to be credible?

Every day, newspapers, magazines, television, radio, blogs, podcasts, and websites disseminate millions of news stories. Although each has a varying level of credibility, we have few means to measure quality. As a result, an increasingly frustrated public struggles daily to judge the accuracy, fairness and integrity of news reports, individual reporters and media organizations. More than ever, citizens need reliable tools to find news they can trust and to make informed decisions about issues important to their lives. So what are we to do?

My Shorenstein fellowship presented me with an opportunity to examine the crisis of confidence in media. "How can we find credible news and information we can trust?" I had asked when applying. "Are there any journals and journalists that we can really trust and rely on? If so, how can we possibly find them amidst the clangor and the clutter of 'TMI' too much information?"

At Harvard I set forth to examine the idea that one answer to the news media's trust dilemma might be found within the digital information revolution itself—more specifically, within emerging, interactive innovations like social networks. "Are there any practical solutions available?" I wondered. At least one leading media theorist, Arizona State University professor of journalism Dan Gillmor, seemed optimistic: The World Wide Web, Gillmor has written, presents us with "an avalanche of tools and ideas that have enormous potential to make journalism more diverse—and better." After all, as with many other aspects of the Web, both the reporting and distribution of news and information were already becoming highly collaborative activities.

The rise of social media and its increasing importance for journalism and journalists already seemed apparent to me in 2008, but it was still a matter of controversy within the profession, as well as within academic circles. Moreover, social media's own uses and overall credibility still seemed to raise many questions at the time: what are the best filters for coping with the flood of information that threatens to engulf us? Can social platforms and their online toolkits really offer valuable help in evaluating news and information?

Are brands still reliable or are they fast becoming obsolete? Will algorithms and recommender systems, "learning machines" that can ascertain our needs and preferences over time, ever learn to fulfill our needs as well as our preferences? What roles might individual content curation, online crowdsourcing, and other human interventions play? Can they enable us to discriminate between good and bad journalism, or do they simply mirror popularity and reinforce prejudice in consciously created communities of likeminded "users?"

In essence, I wanted to explore whether or not social media such as Facebook, Twitter, YouTube and the like in fact could supply a missing piece to the puzzle that was journalism's trust problem by helping us to leverage technology and find credible, high-quality news and information. There had never been a better or more important time to find out than in Fall 2008, especially with a crucial and historic national election looming. "At the moment, there still are more questions than answers," my research proposal had concluded, "but as Gillmor notes, this 'nears holy-grail territory . . . for sorting out the good from the bad, the useful from the trivial, the trustworthy from the phony.'"

I was already convinced that the rapid rise of newly emerging communications technologies signaled a huge shift in how we find, consume and interact with news and information of all types. Three clear trends seemed most evident: first, it was becoming more difficult than ever to separate fact from fiction and truth from spin in any form of media, old *or* new, legacy *and* emerging; second, individuals and institutions were increasingly using social media to create, curate, and transmit news and information of all types and origin; and third, facing an unprecedented flood of content, we wanted—and desperately needed—a variety of new information filters to assist us in separating the signal from the noise, the real from the fake and the trustworthy from the downright incredible.

Trust . . . But Verify

Although there was widespread consensus that new information filters were needed to help us navigate through the flood of information now washing over us in real time and to separate signal from noise, there was little agreement as to what might be the best way to filter. Some leading academic researchers and technologists were adamant in their belief that algorithms, learning machines, and recommender systems would become sophisticated enough to provide us with both personalized news that we choose *and* serendipitous news we can use; others looked instead to informed and informative "tastemakers" and "influentials" whom they hoped would take on the roles of trusted content curators, valued editors and even fact checkers, perhaps even to evolve over time into "micro-brands" themselves.

Corporate executives, including many legacy media leaders but also the likes of former Google chairman and chief executive officer Eric Schmidt, were clinging instead to brand power as the answer. Schmidt denounced the Internet itself as a cesspool where false information thrives. "Brands are how you sort out the cesspool," he maintained. "Brands are the solution, not the problem."

Other Internet luminaries disagree strongly with Schmidt; Craigslist founder Craig Newmark, who has termed trust "the new black," is but one prominent example. But though Schmidt may have overstated his case—denouncing the Internet as a cesspool of misinformation is akin to blaming the telephone system for any and all lies people may tell while using it—addressing the Internet's trust issues is certainly central to its future development across a broad spectrum of applications and industries.

Predictably, Internet trust concerns such as those expressed by Schmidt were most evident in the news arena and widely shared by traditional journalists.

While concerns over the credibility of online news and information were certainly valid, however, I was puzzled as to the overwhelming focus placed on "Internet news" to the exclusion of the legacy media. After all, evidence of misinformation, disinformation, inaccurate reporting, fake news, phony news releases, pay-for-play punditry, and a host of other media malpractices was equally prevalent offline as online, if not more so. As a result, public confidence in the traditional news media had been declining for years and the rate of decline was accelerating.

Gallup polls throughout the 1970's, for example, showed that about 70 percent of those surveyed had either "a great deal or a fair amount of trust and confidence in the mass media—such as newspapers, TV, and radio—when it comes to reporting the news fully, accurately, and fairly." But by 2004, the percentage had flipped: less than one of three Americans expressed a "great deal or a lot" of confidence in newspapers and broadcast news. Just four years later the Gallup trust figure dropped to fewer than one in four. A Zogby poll later in 2008 also revealed similarly widespread distrust; nearly three-fourths of those surveyed said they believed that the news they read and see is biased and simply not credible. By 2010, the results of the annual Gallup poll were even more dire: a record high number of respondents said they had little or no trust in the mass media to report the news fully, fairly and accurately. Meanwhile, the percentage of those who expressed "a great deal or fair amount of trust" tied the record low. These findings were further confirmed by a separate Gallup poll that same year, which revealed little confidence in newspapers and television specifically.

The "Bottom Line," as Gallup pollsters phrased it, was that the "annual update on trust in the mass media finds Americans' views entrenched," with a record-high number "expressing little to no trust in the media to report the news fully, accurately, and fairly, and 63% perceiving bias in one direction or the other." For what little it was worth, the good news was that the media were still trusted slightly more than Congress, which is traditionally the nation's least trusted institution. The bad news was that they rated lower on the trust scale than the other two branches of government—

even though trust in all parts of the government was itself down sharply.

Although still trusted more than Congress, the traditional news media in general were clearly subject more than ever to a growing lack of public confidence. (Respondents on both the right and left of the political spectrum, who seldom agree on much else these days, seemed to share at least this negative assessment.) Clearly journalism's trust issue was not just a problem for journalists. The breakdown in the relationship between journalists and the "people formerly known as the audience" also presented a serious social challenge. If we cannot ensure that we are receiving credible news and information, the implications for our democracy, which depends on an active, informed citizenry, are enormous.

The Most Essential Medium

Fortunately, there was hope on the horizon. As online platforms like Facebook, Twitter, and YouTube began to fight for attention with what was coming to be known as the mainstream or corporate media, they also began to diminish our longstanding dependence on legacy media of all sorts, most of which could no longer rely on brand power to inspire trust and confidence in their products. At the same time numerous public opinion surveys revealed that the use of new media was rapidly expanding in all demographic groups, although it was predictably highest among the "Digital Natives" aged 18–29, who have grown up in an

Internet-enabled reality and who take as a given this stunning democratization of media, and to whom the very idea that anyone/everyone can have a voice seems self-evident and hardly remarkable. (One recent Time Inc. research study revealed profound differences in the way different generations process media and indicated that the brains of young people may now in fact be "wired" differently.)

In any event, recent Pew Research Center surveys shows that Americans of all ages increasingly get their news from multiple sources. Most respondents say they use Internet-based sources such as websites, blogs, and social networking sites, and only a minority rely entirely on traditional sources, including print, radio, television, and cable news. As one Project for Excellence in Journalism's State of the News Media report noted, "Americans are going online more frequently, spending more time there and relying more on search and links rather than brand-name destinations to navigate the Web." The survey concluded, "The Web is becoming a more integral part of people's lives. Eight in ten Americans 17 and older now say the Internet is a critical source of information." As the survey showed, Americans increasingly identify the Internet as a more important source of information than television, radio, and newspapers, and many now believe the Internet is the "most essential medium."

Politics "2.0"

Although first to feel its effects, the Web-dependent worlds of media and technology were not the only ones to be rocked and roiled by the information revolution. Politics of all sorts were also being drastically transformed, both domestically (see Chapter 11) and internationally (see Chapter 14). Within the United States, for example, beginning in 2003, social media rapidly redefined the very nature of campaigning for national office, as first Democrats like Howard Dean and Barack Obama, and later their Republican rivals, learned to employ these new tools to great effect. As in the related fields of media and technology, traditional notions of how to drive a modern political campaign were upended overnight, particularly those regarding how to raise funds and organize support.

Would this revolution in American campaign politics mean we would next receive our political news and information through direct distribution from politicians? If so, would such information be any more worthy of our trust than that transmitted by journalists and traditional news media? How much of social media's impact on politics is real and how much an admixture of hope and hype on the part of advocates remains a matter of debate, but by 2011 there could no longer be any doubt that emerging social platforms threatened long-established legacy political brands in much the same manner they already had those of the legacy media.

Both within and beyond the borders of the United

States, citizen advocates had already employed social platforms such Facebook, YouTube and Twitter in a number of ways and with varying efficacy, from opposing self-styled revolutionary groups such as Columbia's FARC to supporting change—some say even fomenting revolution—everywhere from Ukraine to Iran. Social media had been lauded in some circles as game-changers and credited with having a revolutionary impact on politics. Perhaps the most vivid example came when the anonymously recorded and distributed video of the death of student Neda Agha-Soltan, shot during protests of Iranian President Mahmoud Ahmadinejad's disputed re-election, was awarded a prestigious Polk journalism prize celebrating "the fact that, in today's world, a brave bystander with a cell phone camera can use video-sharing and social networking sites to deliver news." More recently, the 2011 spring uprisings that toppled authoritarian regimes in Tunisia and Egypt, as well as the global economic protests of the autumn, were coordinated in part on Facebook and Twitter.

Critics contended, however, that there was little real evidence that social media are actually effective as political organizing tools. *New Yorker* writer Malcolm Gladwell sparked much conversation and controversy when he questioned evidence of "Moldova's so-called Twitter Revolution," for example, and Evgeny Morozov, author of *The Net Delusion: The Dark Side of Internet Freedom*, suggested the claims that social media lead to social change were highly suspect and the product of "cyber-utopian" thinking. Instead, Morozov argued, "The Internet has provided so many cheap and easily available entertainment fixes to those living under authoritarianism that it has become considerably harder to get people to care about politics at all." Nevertheless, a variety of repressive regimes, unhappy with the news and information their citizens' share, have regularly shut down not only social platforms like Twitter and YouTube but even the entire Internet itself—though usually not for long.

Moreover, although this overall disruption now seems most prevalent in the context of either politics and governance or news and information, what is affecting these sectors should be regarded as a mere harbinger of what is soon to come in virtually every other arena of societal endeavor, whether individual or institutional, corporate and governmental. Along with such media-related industries as music, news, television, film, publishing, and advertising, other entire segments of the economy including real estate, telecommunications, and finance have already been radically disrupted; healthcare, education, energy, and many others are poised to be next.

"Predicting the future of the Internet is easy: anything it hasn't yet dramatically transformed, it will," as technologist and investor Chris Dixon noted in his blog. "The modern economy runs primarily on information, and the Internet is by orders of magnitude the greatest information mechanism ever invented. In a few years, we'll look back in amazement that in 2012 we still used brokers to help us find houses, that doctors kept records scribbled on notepads, that government information was carefully spoon-fed to a compliant press corps, and that scarcity of information and tools was a primary inhibitor to education."

To paraphrase Internet guru Clay Shirky, "*Here comes everything!*"

2.

Brands, Cesspools, and Credibility

The Internet is a "cesspool" of false information, says former Google Chairman and CEO Eric Schmidt, and the future of credible, trustworthy "quality editorial" is a "huge question in the world, particularly in the United States." Instead of relying on such information filters as algorithms and recommender systems, online social networks, or expert curation, Schmidt remains resolute in his belief that brands are absolutely essential in helping us to navigate the new information world. "Brands are the solution, not the problem. Brands are how you sort out the cesspool," Schmidt says.

"Brand affinity is clearly hard-wired," Schmidt once told a gathering of American magazine editors seeking his oracular advice. This attribute "is so fundamental to human existence that it's not going away," he assured his audience of traditional gatekeepers. "It must have a genetic component."

Although there is widespread agreement on the need for some type of filter for the delivery of credible, trustworthy news and information, there is predictably far less accord as to what the best one may be. Not surprisingly, many executives in legacy media companies still share Schmidt's faith in the fundamental power of their brands. Richard Stengel, executive editor of *Time* magazine, is among them. At a Time Warner "Media Summit" in October 2008, Stengel remarked, "I actually think that in this blizzard-like universe of news usage, brands are actually more important and rising above the chaos because people don't have places they can trust and rely on."

ABC News executive Paul Slavin, Senior Vice President of Digital Media, agrees. "Brands are the answer to the credibility questions," says Slavin.

"ABC News is known worldwide, and most people feel we are balanced and fair, that we offer a vetted, careful environment for news and information." He believes that brands are already being used as a necessary filter to combat the "too much information" problem, and that our reliance on them will actually grow over time. "Brand power will only increase as noise level increases," Slavin explains. "It will all come back to tried and true brands. The fundamental understanding of and protection of our brand truly is our future."

Slavin, who has coordinated ABC News' exploration of and collaboration with the emerging media, said in an interview that ABC first "started looking for relationships with social networks" in early 2007. "We looked at MySpace first, then Facebook," he recalled. The motive, Slavin says, was simple, "We wanted to tap into their younger demo and expose them to our content." In the end, ABC chose to work with Facebook.

"Facebook friends function as personal aggregators and that can be very powerful," Slavin noted. "We needed to figure out how to tap into that."

ABC entered into a formal partnership with Facebook in November 2007, the first such arrangement the social network made with any traditional media outlet. The agreement enabled Facebook users to follow ABC reporters electronically, view reports and video and participate in polls and debates. The companies also announced that they would collaborate to sponsor a presidential debate in New Hampshire to be held in January 2008. Facebook users around the world could connect and instantly discuss the debate as it was broadcast live on ABC.

The ABC Facebook page received a lot of traffic "when they actively promoted it," Slavin recalled. "We had very good cooperation and coordination initially, and it resulted in 1.5 million downloads." The social network later changed direction, however, and decided it didn't want a strong relationship with just one media entity. "We had talked about more collaboration in the general election," Slavin said. "Our goal was to expand our audience to include people not coming to us for news already. The Facebook relationship can be very powerful if and when Facebook wants to do it and pushes it."

Why was ABC so interested in the online social networks? "We were looking for ways to connect with their audience of young people," said Slavin. "In terms of the election, it gave us another way to communicate and to generate interest and questions for town halls. We had already looked at YouTube—then they did a debate with CNN and got very hot." YouTube's partnership with CNN for two debates during the primary season was perhaps the most high-profile emerging media relationship during the 2008 presidential election cycle. Users were able to upload video questions to the candidates, which were then vetted by YouTube and CNN personnel. Those selected were then played during the debates.

ABC News would "love to work with Facebook more," Slavin confessed, and he is "looking to re-engage and expand the relationship." Although he still finds YouTube interesting, he "is not sure what we would get out of the relationship, since there is no money to be made—maybe marketing?" In conclusion, Slavin noted, "Everybody is grappling with this now. This is the most interesting time I've ever experienced in news business. There's such an explosion of new technology that my main problem is that there are simply not enough hours in the day to deal with it all. Everyone is talking to everyone else, and we're all trying to figure this out."

NBC News Wants To Be Your Trusted Friend

Mark Lukasiewicz is another broadcast network news executive who is grappling with the related issues of legacy media, emerging social media and trust. Lukasiewicz, formerly Vice President for Digital Media at NBC News, takes issue with much of what both Schmidt and Slavin have to say. "The

Internet is a conveyor belt for information, not a repository of it," says Lukasiewicz. "Let's not blame the messenger. The Net is no more of a cesspool than life in general.

"The question is: What tools do people have to determine what is true?" Lukasiewicz says. "In previous times, the medium itself conveyed some of that trust relationship. But now, since so much information comes through this new device of the Internet, it's become a lot harder to make those distinctions. Branding is *part* of what's necessary," he believes. "But the big challenge for mainstream media like us is that people today are less trusting of news brands—the war in Iraq had a great deal to do with that—and now this new ability of people to find and share information on their own feeds into that."

Lukasiewicz says that other fundamental changes are also shaking the firmament of the legacy media. "After all, what conveys authority?" he asks. "That is what is changing. Today, for us in the mainstream media, being a singular provider—the *one brand*, offering everything it and only it produces—is actually a negative. People want to see a multiplicity of sources; they want you to be comprehensive. So if we link out, and offer content other than our own, even that of our competitors, it still enhances our own brand in the eyes of the consumer.

"I know there is more than one vision and more than one view point," says Lukasiewicz. "Multiple view points are what consumers want. So you build a trusted brand by sharing others' content. It sounds a bit paradoxical but it works, even though in traditional media terms, linking to the competition once would have been seen as self-destructive."

Lukasiewicz thinks that creating some sort of hybrid social/brand filter may be the key. "The brand that increasingly matters is the one called 'my friend,'" he says. "People don't *come* to Facebook for news content, but they *get* it there. So yes, NBC News wants to be your trusted friend. And I do that by being in all the places where you are—cell phone, online, in the back of a taxicab, on a screen in an airliner, on Facebook, you name it—when I do what I do. I want to be there for you, where and when you want it.

"All this is still rapidly evolving," Lukasiewicz concludes. "For credible coverage of major events, for example, people still turn to trusted news brands. But in the future, I really believe that if you can create a cross-platform home for your news delivery, you will also succeed in creating a trusted brand."

Today's brand leaders, Lukasiewicz believes, must embrace social media or risk rapidly ceding both audience and relevance to a variety of competitors from outside the mainstream. Media consultant Terry Heaton once described the phenomenon on his blog, "The speed with which a media brand can be built out—see Huffington Post for the most breathtaking example—means that the barriers to entry that made the media business the province of titans are gone. On a journalistic level, the new playing field is more even. Many people see the news in aggregated form on the Web, and when they notice a link that interests them, they click on it with nary a thought about the news organization behind it. Information stands or falls on its magnetism, with brand pedigree becoming secondary."

Consider the recent sales, for example, of just two media companies, Newsweek and TechCrunch. Investor Sidney Harman paid just one dollar for *Newsweek*, a leading print magazine with a trusted brand built up over decades; AOL purchased TechCrunch, a five-year old technology news and aggregation site that began as a one-person blog, for a reported twenty-five million dollars. As Heaton explained, "TechCrunch IS Michael Arrington, beginning *just five years ago* as his blog. It carries his persona, and the views of its writers are understood by those who make up its audience. . . . We've entered an era of personal branding and argument-laden prose that helps people figure out life around them through their surrogates, the men and women who bare their lives and views for all to judge. The idea of protecting a media brand . . . is becoming less and less practical."

At the same time, as per Lukasiewicz, the dichotomy between mainstream and digital media is rapidly disappearing. "Formerly clear bright lines are being erased all over the place. Open up Gawker, CNN, NPR, and the *Wall Street Journal* on an iPad and tell me without looking at the name which is a blog, a television brand, a radio network, a newspaper," *New York Times* media columnist David Carr challenged his readers. "They all have text, links, video and pictures. The new frame around content is changing how people see and interact with the picture in the middle. . . . So if news is wherever the public finds it, what really is the value of creating a complicated, labor-intensive print product?" If the *Newsweek* sale is any indication, in some cases that value is now nearing zero.

Rafat Ali, founder of paidcontent.org, a leading industry-monitoring site, is another close observer who says that media news brands are severely threatened. "The only constant in the news industry is change," Ali said in a 2010 interview at the Poynter Institute. "And how do you build a lasting brand, a business in that kind of sector? There's too much undercutting. The fact of life in the news industry is everybody undercuts everybody. Everything undercuts everything. Every new technology undercuts every technology, which will then get undercut by a new technology that will come along."

Even powerful new brands like Google, Facebook, and Twitter are not "immune to the creeping forces of creative destruction," as social media strategist Gini Dietrich warned in a blog post on the socialmediatoday.com site. "AOL, Friendster, and MySpace were once considered inexorable juggernauts," Dietrich pointed out. "Over time they all fell short of user expectations in various ways, allowing alternative value propositions, service models, and providers to gain a foothold and eventually surpass them."

Companies like Facebook and Twitter, Dietrich argued, "rose to their preeminent status by capitalizing on the shortcomings of their predecessors, then relying on the difficulty of moving/replicating personal networks to keep users in place." But at the very same time as these "new hegemons enjoy seemingly limitless growth," Dietrich said, they may ultimately also be sowing the seeds of their own demise.

Such seeds of discontent that could destroy the new media brands include the continuing erosion of privacy (Facebook's approach was once described as step on toes until people scream, then apologize); a lack of sufficient controls over personal information and content; poor usability; and what Dietrich terms "the classic," hubris. "Today's social media mega-networks," she reminds us, "appear to underestimate the power of a highly skilled, highly motivated community of true believers to effect transformational change over time."

There are already several examples of new approaches that might one day topple the new media brands. Location-based social media such as Foursquare could revolutionize advertising models and dethrone Google and its less targeted approach, for example. Or open-source social networks like Diaspora could one day become what its creators promised, a "social network that puts you in control of your information," where "you decide what you'd like to share, and with whom," and where you retain full ownership of all your information. New, niche private networks or social customer relations management (CRM) services such as Gist, which "let individuals manage and interact with their personal and professional networks by aggregating contacts from various sources in a single place and

LIVERPOOL JOHN MOORES UNIVERSITY LEARNING SERVICES

providing tools for content sharing and listening," could also become popular.

With legacy media brands in free fall, and new social ones ascendant, what does the digital disruption promise for other industries and brands? Richard Sambook, who spent decades working at the BBC, including a four-year stint running BBC News, says non-media brands must now begin to tell their own story and become content creators themselves, rather than relying on any intermediary.

Sambrook began thinking about the changing relationship between media makers and the public in 2004, while still at the BBC. "I began exploring social media and the role it was playing and how it was transforming relationship between the public and organizations and traditional media," he explained to Fast Company blogger Steve Rosenbaum.

Now Chief Content Officer for Edelman Public Relations, the largest independent public relations firm in the world, Sambrook's new job is to work with global clients to develop their own content. He advises them that they must both talk and listen in this new social media world. "Brands have to learn humility in order to get some traction in that conversation," Sambrook concludes.

Roger Fransecky, CEO of the corporate-consulting firm Apogee Group, agrees. In an age of "digital distraction," he argues, "The powerful metaphor going forward is 'conversation,' and media brands are no longer information providers." Instead, Fransecky tells them, "You're in the conversation business." From a branding perspective, he believes, the challenge is to ask a series of questions. "When you look at your business, you need to ask, 'what do we still trust?'" he concludes.

Peer Influencers

Facebook, Twitter, LinkedIn, YouTube and many other social platforms are now deeply intertwined with both our personal and professional lives. Yet these platforms are only "the igniting technologies" of a social media revolution now in full swing. Today, consumers determine the news and topics that hold their interest, and they rely primarily on social network conversations—even more than search engines like Google's—to filter that news.

New technologies allow us to monitor what is being said amid the rising noise level on the social Web and to help identify "influencers" who act as filters, helping to shape conversations and reputations. Such influencers have previously been called "Mass Connectors and Mass Mavens," first by *New Yorker* writer Malcolm Gladwell in *The Tipping Point*, and later by Augie Ray and Josh Bernoff in a 2010 Forrester Research study, *Peer Influencer Analysis*.

"Influence comes in two types," Ray and Bernoff explained. "First, there is influence from people posting within social networks: Facebook, Twitter, MySpace and so on. We call these instances influence impressions. Second, there is influence created by posts: blog posts, blog comments, discussion forum posts, and ratings and reviews. We call these influence posts."

The Forrester team estimated that people in the United States alone create 256 billion influence impressions on each other in social networks every year. Of these "influence impressions," 62 percent come from Facebook. They also estimated an additional 1.64 billion "influence posts" every year—another 250 billion-plus impressions. Add the two—impressions from both social networks and blogs together—and you get more than 500 billion impressions *made by people influencing their peers.*

From their analysis of this huge pool of influence, the researchers concluded that people's influence on each other already rivals online advertising by brands—and that such peer impressions from "mass connectors and mass mavens, aka influencers, are much more credible than most advertising, since they come from friends. It's time to start analyzing peer influence with the same discipline we apply to media," they concluded. "We are just at the beginning of understanding the best uses for this tool."

What might the future of influence filtering hold? For one thing, we can expect to see a new generation of social media monitoring and engagement tools. These might include databases containing the universe of influencers and the archived content they generate, real-time monitoring of online content, "with automated text and sentiment analysis capabilities that give a fuller understanding of online conversations and the profiles of influencers," as the Forrester research team out it, and real-time measurement and analysis of social media communications. Contextual search and queries will enable relevant questions, accurate answers and increased credibility of news and information found both online and off. Influencers will build out their reputations as expert sources and eventually become more trusted and useful—even ending up as information micro-brands in their own right.

Information will also become far more dynamic, as the influencers discuss topics in real-time; tagging engines will offer the content of previous articles or posts authored by that influencer, and a tag cloud of subjects or keywords will replace journalistic beats. Machine-based "sentiment analysis" will begin to predict how influencers may cover a topic. Our social search to separate signal from noise, credible from incredible, will be enhanced by understanding how and why both brands and peers exert influence.

No Dominant Players, No Dominant Brands

It's become a truism that everyone is now responsible for his or her "own brand." But within the ongoing social media revolution, with its learning machines and recommender systems, human networks, peer influencers and old and new media firms battling for attention and relevancy, what does the word "brand" even mean? What if the future of media is that there will be no dominant players at all and no dominant brands—at least as we now understand the term?

In a blog post that examined Gawker Media and its founder, headlined *Is Nick Denton Really the New Rupert Murdoch?*, the *New Yorker*'s John Cassidy asked the following questions, "Can Gawker Media (and other blogging outfits such as the Huffington Post) translate their rapid audience growth into big streams of revenue and profits, thereby becoming dominant players in the news-media business? Or will the established players, which now have sizeable online arms as well as other sources of income (and costs), ultimately come out on top?"

"This kind of thinking drives me nuts," Scott Rosenberg, a founder of the online magazine Salon, responded in a blog post. "It's always a zero-sum battle for dominance. (Can the scrappy little new guys grow so powerful that they'll replace the big old guys? Or will the lumbering big old guys survive and 'ultimately come out on top'?) And it always misses the point."

Rosenberg doesn't believe that Denton's Gawker, AOL/Arianna Huffington's Huffington Post, and similar-scale ventures will ever become dominant players. Instead, he says, "those that husband their resources and play their cards smartly will survive, continuing to grow and to figure out the contours of the new media we are all building. They'll be active, important players, without 'dominating' the way the winners of previous era's media wars did."

The legacy media, what John Cassidy calls the "established players," will then fall into two groups. "Many will collapse under the weight of their legacy costs and dwindling revenues, as so many are already starting to. Others will survive by figuring out, in time, how to cut costs while expanding their online reach," says Rosenberg. "The survivors in the second group will find that they can be profitable and do good work, but they will hardly have 'come out on top.' In fact, as companies, they will come out looking much more like Gawker Media and Huffington Post than today's Time Inc. or New York Times Company."

Of course, there is also a possibility that new dominant players may enter the arena as well. "Just look at the investments AOL and Yahoo are making in content," Rosenberg notes. "But I think they'll find dominance elusive, too." In other words, the media future will most likely see no small group of dominant players, aka brands. Instead there will be "a much broader spectrum of modestly successful players," says Rosenberg. "This is because, in a world awash in content, the media business is never going to be as profitable as it was in a world of scarce content. It will be sustainable, but it won't support the sort of monopoly profits that made it so attractive for seekers after dominance."

Meanwhile, as media outlets increasingly begin to look like loose federations of individual brands, the same forces of fragmentation will apply. Asked to name the journalist or newsperson they most admire, for example, half the American public now offers no specific answer, and no journalist is named by more than 5 percent of the respondents. With recent media consumption surveys showing that nearly as many Americans now get news from both traditional and digital platforms as from traditional platforms alone, the decline in mentions of admired journalists—individual brands, if you will—is apparently a response to a much wider array of news choices than in earlier eras. One study by The Pew Research Center for the People & the Press, for example, showed that in 1985, nearly two-thirds could name a favorite journalist; today, 52 percent cannot name anyone.

3.
Can Brands Be Trusted?

When longtime Google CEO Eric Schmidt called the Internet a cesspool he put forth well-known corporate brands as the best filters, necessary to help us sort through all the muck and mire of false information found on the Internet. But leading researchers who have looked into online credibility issues in depth, such as Eszter Hargittai, associate professor at the Department of Communication Studies at Northwestern University, have begun examining the filtering role that brands already play, and have come to some surprising—and in some cases downright scary—conclusions about their effect.

The Internet, as Hargittai notes, "is a source of unprecedented amounts of content . . . both lauded for its breadth and critiqued for its sometimes free-for-all ethos." In this information-rich environment, however, "traditional gatekeepers such as editors no longer evaluate material before it has the potential to reach large audiences."

As legacy media brands began cutting back or even

in some cases entirely eliminating their previous efforts to "evaluate material before it has the potential to reach large audiences," Hargittai found that the ability to find trustworthy content online had become "an essential skill." Those seeking reliable news and information still look for shortcuts and filters to assist them—and they still rely on certain brands, as Schmidt noted. But the most trusted brands are no longer found among legacy media enterprises, such as the major magazines to whose editors he extolled brand power. Instead, as Hargittai's research shows, "users put considerable trust in the online equivalent of traditional gatekeepers: search engines."

In the first decade of the twenty-first century, search engines began to replace more traditional intermediaries as a means of finding trusted content. Small wonder then that Eric Schmidt is so enamored of brands—with Google leading search, it was the Google brand itself that soon became pre-eminent. "Search engine use is one of the most popular online activities," as Hargittai and her associates noted in one study, called *Trusting the Web: How Young Adults Evaluate Online Content*. According to the report, nearly half of all Americans using the Internet "turn to a search engine on a typical day," the figure is even higher among the young adults surveyed, and two thirds believe that "using search engines provides them with 'a fair and unbiased source of information.'" Search engines are now a "crucial part of the puzzle of online credibility assessment. . . . They have become the most prevalent tool for information seeking online *with the potential to garner large influence . . . on what material users deem trustworthy.*" [Italics mine]

Hargittai's research showed that brands were "a ubiquitous element throughout our respondents' information-gathering process." But it also revealed a frightening lack of knowledge as to how brands such as Google actually operate in the information sphere. The study noted, for example, that only 38 percent of Internet users were aware that sponsors paid for their links to appear first on Google's search engine results page. "Our findings suggest that students rely greatly on search engine brands to guide them to what they then perceive as credible material simply due to the fact that the destination page rose to the top of the results listings of their beloved search engine." Google's branding is so powerful, in fact, that more than a third of the study's participants used its brand name as a verb, regularly responding "I'll google it" when asked how they would complete an information-seeking task—despite the fact that the company admittedly performs no credibility verification whatsoever of the information links it offers and features paid sponsored links more prominently than others.

The effect of branding is strong enough, especially among the young, that almost all (98 percent) participants in the study sample mentioned a name brand at some point. Google (85 percent) and Yahoo (51 percent) were mentioned most frequently, followed by several other leading brands, including Facebook. "Known brands were essential signifiers of quality for respondents, and seem to serve as an important part of users' daily information-gathering routines," the study notes. "Mentions of corporate brands dominated students' reported habits, with 63 percent of all respondents mentioning a corporate brand as part of their routine search behavior."

These findings suggest to Hargittai's team that, "while users may feel confident in their ability to find accurate and credible information online, that confidence may not be translating into an increased skill level in credibility assessment." In other, less academic words, this means that although users *believe* they can find trustworthy news and information online, that belief is mistaken. Perhaps worse, "students' level of faith in their search engine of choice is so high that they do not feel the need to verify for themselves who authored the pages they view or what their qualifications might be."

"We Don't Dominate the Conversation"

Despite the claims of Google's chief and other leading executives such as *Time*'s Richard Stengel or Paul Slavin of *ABC News*, corporate branding—whether from traditional media gatekeepers or their modern counterparts in search, such as Google—is clearly insufficient in solving the ongoing credibility dilemma. While reliance on such trusted brands might have provided a partial answer in the past, social platforms now supplement and in many cases even supplant search engines in the quest for reliable information. At the same time, the longstanding power and reach of news media brands in particular are quickly diminishing.

The world now "has many, many places to turn for information, misinformation, analysis, rants, etc," as then-*New York Times* editor Bill Keller wrote, decrying this trend in September 2008. "We—*The Times*, *The Washington Post*, Politico, the news outlets that aim to be aggressive, serious and impartial—don't dominate the conversation the way we once did, and that's fine, except it means some excellent hard work gets a little muffled . . . I've been repeatedly surprised at the rich, important stories that fail to resonate the way they deserve."

Keller noted, "On one level, more people read the *Times*, albeit in digital form, than ever." Important *Times* articles about the recent presidential campaign, he added, "did a brisk business as an e-mail forward. But so did everything else anyone had to say that day about the campaign—*whether it was true or false, reported or simply asserted, fact or opinion* . . . [Italics mine] Everything is equal, everything is a tie and nothing, it seems, is important anymore.

"Nobody has felt this more acutely than the Newspapers and Magazines of Record in the United States," Keller concluded. "*The New York Times*, *The Washington Post*, *Time*: all over the world of 'quality' journalism, there is a feeling of decline."

Bill Keller is right—the old, self-referential and certainly self-defined world of branded, so-called "quality" journalism is in deep decline. Yet at the same time, engaged citizens continue to seek out true quality and credibility in a world that now indeed has many, many places to turn for information—and misinformation. It is unsurprising that the use of social media for the delivery of news and information is rapidly increasing in all demographic groups, although adopted most quickly and widely by the young. Studies by a new breed of media researchers (see Chapter 9) provide ample evidence that people continue to rely on those in their networks when seeking various types of information, and that emerging social media can and do play a useful role in helping us to access reliable, credible and trustworthy news.

More research and empirical evidence is still needed, however. One reason is that relatively few studies have yet to look at what users actually do to assess online credibility. As Hargittai notes, "While considerable prior research has examined what users *claim* to do in order to find credible information online, research that compares *actual* and reported behavior is less common."

Filters and Shortcuts
Ultimately, however, the question remains: given the plethora of information now widely and readily available, are average citizens really interested and capable enough to decode that which is useful, credible, quality information—and that which is not? Even if interested and capable, will they take the time necessary to do so? Most careful observers agree

that some filter or shortcut is needed to assist us in sifting through the overload of information. As communications researcher Miriam Metzger, associate professor at the Department of Communication at the University of California, Santa Barbara, says, "People know they 'should' critically analyze the information they obtain online, yet rarely have the time or energy to do it. Most current research shows people want to use shortcuts in determining trust and credibility. This is something known as known as 'credibility heuristics'—a kind of information Verisign, if you will."

"Only the truly motivated will actually do the work required," Metzger concluded. "The rest of us need and want filters. Can social networks play this role? If so, will filtering best take place in already trusted environments like Facebook? It certainly makes good sense to me—in terms of credibility at least."

"People are always looking for trust shortcuts," agrees Kelly Garrett, assistant professor at the School of Communication at Ohio State University. "It's either brands, some sort of credential, or some sort of social network—but they are making up their own ways of trust assessment." Stanford University's BJ Fogg also believes, "Brands can be shortcuts," but points out they are quickly losing prominence.

"The mainstream media *had* a sort of trusted brand but they've given up a lot of trust of late," says Fogg. "The issue around brands is that different friends trust different brands. The challenge now is that there are no destination sites—so that undercuts the value of news brands. And lost trust equals a lost brand." Fogg concludes that the legacy media "deserve what has happened to them, and once you lose credibility, it's very, very hard to regain. It's difficult to change people's habits—especially the young—once that trust and that brand is damaged."

4.

The Facebook Decade: F8 = Fate?

"Imagine . . . that you knew which sites—or what news stories—people you trust found useful and which they disliked," David Kirkpatrick wrote in the June 11, 2007 issue of *Fortune* magazine. "This isn't fantasy. Facebook might make it possible, and soon. Yes, the social-networking site college kids spend so much time on—the one you thought was just about hooking up—could turn out to be more important than any of us thought."

Kirkpatrick, who was then *Fortune*'s Senior Editor for Internet and Technology, went on to write the best-selling *The Facebook Effect: The Inside Story of the Company That is Connecting the World*, the definitive book on the company. He was prescient. In a startlingly short period of time, Facebook *did* make it possible for you to find those trusted and useful news sites and stories—along with much, much more.

An online social network founded in a dorm room at Harvard University in February 2004, initially available exclusively to people affiliated with Harvard, Facebook soon opened its membership to people associated with other colleges and universities, and then to those in high schools. In September 2006, two other key changes followed: first, an automated flow of information sharing about activities and interests among others in your social circle, called the "News Feed," was unveiled; second, Facebook membership was opened to everyone.

Opening to the world led to immediate and spectacular growth for the still-nascent company. Traffic to the Facebook website increased 26 percent in just three months, while the News Feed gave it unprecedented power to disrupt further an already beleaguered mainstream media. Rather than having to search for trustworthy news and information, Facebook users now had it delivered directly to them in real time. "News Feed brought interesting things to people's attention quicker, so they looked at more content," Facebook co-founder and CEO Mark Zuckerberg explained to Kirkpatrick.

Despite its rapid and scalable growth, however, Facebook's future remained uncertain. Although Zuckerberg had impressive plans, he claimed his creation would become "the most powerful distribution mechanism that's been created in a generation," Facebook also faced major challenges when David Kirkpatrick gazed into Fortune's crystal ball in 2007. For one thing, its CEO was a callow and still largely untested 23-year-old who "sometimes lapses into jarringly grandiose language," as Kirkpatrick phrased it; for another, despite having already grown to 24 million members, less than half of whom were in college, Facebook remained just one among many hopefuls in the highly competitive and fragmented online social networking space. In April of that year, for example, although Facebook counted 14 million unique U.S. web visitors to its site, the leading social network MySpace had more than four times as many. MySpace was also profitable and flush with both cash and possibility as a result of its nine-figure purchase two years earlier by media titan Rupert Murdoch.

Ever confident—some might say cocky—Zuckerberg remained unfazed by MySpace's dominance, even though it was the leading website on the Internet as measured by monthly page views. "We just have such a different philosophy and view of the world," he said. "We're a technology company. MySpace is a media company, and they view their job as owning and distributing content." Zuckerberg viewed his mission quite differently: he planned to unite the world through what he termed a "social graph" comprised of Facebook members and their connections, and then to distribute all sorts of news and information through those connections, which he saw as "the core value of Facebook." Instead of owning and distributing content, he would turn his network into "something of an operating system," he told Kirkpatrick, a platform (or software environment) for many different computer applications, all of which would have a social basis. These applications would be created not by Facebook employees but by outsiders, developers Zuckerberg would invite to take advantage of Facebook's powerful new distribution network.

In May 2007, Zuckerberg announced the details of this plan to hundreds of software developers, analysts, industry leaders and journalists. Facebook would open its network for those interested in building social computing applications—an amazing opportunity for any company or individual software maker to build services for Facebook's members. Zuckerberg said his network was about to become a platform where one could include "friends" in virtually every online activity imaginable

"Today, together, we're going to start a movement," he told the assembled audience. That movement, as journalist Julia Angwin recalled in her book *Stealing MySpace: The Battle to Control the Most Popular Website in America*, centered on "the release of Facebook's application programming interface, or API—a guidebook that enabled developers to write programs that would run on Facebook's website."

"Imagine all the things we're going to be able to build together," Zuckerberg told the crowd at the gathering, which he had dubbed, with typical chutzpah, "F8," pronounced "fate."

BJ Fogg was among the invitees. Fogg splits his time between Stanford University, where he teaches and also directs research and design at the Persuasive Technology Lab, and Silicon Valley, where he works on industry innovation. He has been called by *Fortune* "one of the most sought-after thinkers" in the Valley. Author of *Persuasive Technology: Using Computers to Change What We Think and Do*, and co-editor of the *The Psychology of Facebook: Persuasion in a Social Network*,

Fogg has a longstanding and active interest and expertise in issues of online trust and persuasion.

"When Facebook opened up its API to partner developers, it was an instant game-changer," Fogg recalls. "Although I was intrigued, I hadn't really 'gotten' Facebook until then. After all, it was just three years old and only had 24 million users at the time . . . But once it became a platform, I had that 'Eureka!' moment almost immediately.

"Of course!" Fogg remembers saying to himself. "This is how Facebook just won the game!" He was not alone in his assessment. Many others shared his enthusiasm, including top people at places like Microsoft and the *Washington Post*, which were among sixty-five companies to launch Facebook applications at F8—a clear signal that zeal for social networking was not limited to Zuckerberg and his fellow Facebook executives. Then-chief operating officer Owen Van Natta described the opportunity this way, "Take anything today on the Internet and overlay a lens that is the people you know and trust."

Social Challenges Search

"Social" was about to challenge "search" for Web supremacy. By opening its API to outside developers and freely providing them with a platform and access to its members, Facebook would overcome the prevalent fragmentation and establish its primacy within the social networking world. The developers, who would not be charged for the right to operate within Facebook, were allowed to make money anyway they could—through ads, sales, sponsorships, whatever they could dream up; Facebook would generate most of *its* revenue from advertising. Its leading business partner Microsoft, which was already brokering banner ads for a minimum of $100 million per year, announced at F8 that it would create new tools to foster links between Microsoft's Windows applications and Facebook's network. The decision was transformative for the fledgling company. F8 and the fateful decisions to create a News Feed and open up to the world were indeed real game changers, as BJ Fogg had foreseen.

Within days it was apparent that F8 had been hugely successful. Many of the applications that developers had launched at the event were already struggling to keep up with massive demand flooding in from Facebook users. Just 24 hours after the conference ended, 40,000 people had installed a music discovery application called iLike. Two days later, the number reached 400,000. Another popular app went from zero to 850,000 users in three days. Developers hurriedly began seeking more servers to handle their increased load.

Two weeks later, Facebook's app feeding frenzy and its attendant gold rush were the talk of the tech world. As CNET News staff writer Caroline McCarthy explained, "Now that third-party companies and developers can create custom applications for Facebook members to add to their profiles, building 'Facebook apps' has become a top priority for many Web companies—particularly smaller ones

continued on page 70

Interview with Randi Zuckerberg

Until August 2011, Randi Zuckerberg, sister to founder Mark Zuckerberg, was Facebook's director of marketing. She regularly interacted with media organizations to discuss ways they could partner with Facebook.

ROC: With slumping public approval, journalism is facing a crisis of trust. We're looking at how people can find and share credible news and information in hopes of regaining this trust. Do you think Facebook plays a role in this process at all? If so, how?

RZ: The concept of "the trusted referral" is integral to the success of content sharing on Facebook. We've found that it is tremendously more powerful to get a piece of content—an article, a news clip, a video, etc—from a friend, and it makes you much more likely to watch, read, and engage with the content.

People will always want to consume content from experts and they will always look to trusted news sources and journalists for important news and current events, but the market has become so oversaturated that it is now just as important to rely on one's friends to help filter the news. When you get a news clip from a friend, they are

putting their own personal brand on the line, saying "I recommend THIS piece of content to you out of all the content that is out there,"—just as they would recommend a restaurant, or a movie.

We are beginning to see journalists and news/broadcast companies creating a significant presence on Facebook to engage with Facebook users and help facilitate this notion of the trusted referral to assist with the viral spread of content. When journalists can really engage with this audience and enlist Facebook users to market and share their content, that is such a powerful way to share credible news and information and tap into the implicit trust that people have with their friends.

ROC: The conventional wisdom in academia is that social networks do the opposite; they serve as polarizing echo chambers where users reinforce their own views rather than being persuaded to listen and perhaps agree with others. Why or why not does Facebook fit this mold?

RZ: This is a great question. I think this greatly depends on where you look within a social website. If you are looking at a user profile, you'd probably be correct in that people use that real estate on the site to build their own personal brand. They post photos of themselves, write about their viewpoints, and tell their friends what they are doing and what they are thinking. So yes, if you look at only the profile, you might believe that social media is just a place for a one-sided posting of information about oneself.

However, if you only looked at the profile, you'd be ignoring a tremendous amount of activity that takes place, on Facebook and other sites. Facebook users join groups to discuss issues, topics, and activities that are important to them. They become "fans" of celebrities, brands, public figures, and businesses. They use applications to see photos of their friends traveling the world, read their friends' blog posts, and keep up to date with news and content.

And most importantly, people use Facebook to learn new things about their friends and the world around them. Our mission as a company is to encourage people to share information that is important to them with their friends. Through the news feed on a user's homepage, Facebook users see what their friends are doing, thinking, and talking about. They discover new books, new articles, new videos, new places to visit, and new people to become friends with.

I can't even begin to tell you how many new things I have personally discovered through Facebook and how my Facebook friends have broadened my horizons and introduced me to new things I never would have discovered before. On many days, I hear about the current events because my Facebook friends will post articles and write thoughts about it . . . even before I discover it from a news site. I have discovered new places in the world to visit, have been introduced to new and incredible people, have discovered new music and bands to follow, and have had my views challenged on everything from politics to taste in Broadway musicals.

ROC: Journalists are using Facebook in unanticipated ways. What are some of the main trends you have noticed? Are you surprised at these novel applications? Can you give us details about your interaction with legacy brands like ABC in the past and where you hope to take things in the future? What has your interaction been with other media outlets and individual journalists?

RZ: I think journalists are only beginning to discover what a powerful tool Facebook can be for their content. In my discussions with many mainstream media companies, I constantly hear them talk about why they are squeamish about posting their content on other sites—their content is their lifeblood, it's all they have . . . why would they give it away for free on other sites?

However, I see more and more media companies understanding the importance of allowing people to consume content anywhere they want to consume it on the web, not just at the media company's website. As I mentioned before, I don't think expert journalism will go away—people will always want a trusted, expert opinion when it comes to news, politics, current events, and important topics—but people would rather get that content on a site they are already on, like Facebook, rather than traveling off to another site if they are already on Facebook engaging with friends and doing other things.

When we worked with ABC on the presidential primary debates, we built a really powerful tool together in the "US Politics Application." In this area on Facebook, we allowed users to consume ABC News content and set up special pages for the reporters who were on the campaign trails where they could blog about their experiences and engage with Facebook users. We also strove to make this area extremely interactive, by turning almost every article, piece of content, and question into a "debate/discussion topic" where Facebook users could post their viewpoint and see what all of their friends thought about a specific issue. This information helped power some of the pundit commentary for a high profile, televised primetime presidential primary debate for the New Hampshire primary.

Understanding that there is still a struggle in which media companies prefer to keep their content on their own site, we launched a product called Facebook Connect, which allows companies to incorporate Facebook's social tools into their website. Facebook users can log into other sites with their Facebook login and see what content their friends are consuming and activity their friends are taking on that site. Companies like CNN and CBS have done a great job implementing Connect and that was clearly only the beginning.

ROC: Do you agree that Facebook is increasingly becoming a sort of conveyor belt for the mainstream media's news products? Do you have metrics showing how often and what type of news stories are posted and disseminated on Facebook?

RZ: I would agree with your initial question. We have an incredible tool called Lexicon, which shows trends and insights into what Facebook users are talking about. Around the presidential election, it was fascinating to look at terms such as "Obama," "Palin," "voting" . . . even "Tina Fey!" to see trends in Facebook user discussion as election day got closer and closer. Lexicon allows you to look at the buzz around a certain word or topic on Facebook, and even allows you to drill down to see exactly where in the United States people are most talking about that topic. As this data becomes more and more refined, I think you will start to see this becoming a really powerful way to show the type of news that is posted and shared through Facebook and how often Facebook users are discussing certain topics.

looking to make it big by capitalizing on Facebook's large and loyal user base." More than 40,000 developers asked to be part of the project and approximately 1,500 applications had been produced. "This is unprecedented in the history of the Internet," Facebook's director of platform Dave Morin told developers at a subsequent conference. Within two months they had built thousands of applications and venture capital firms were eager to fund more. In opening up, Facebook was also "locking up the loyalties of the development community, as Angwin phrased it; the strategy was "straight out of the Silicon Valley playbook."

Meanwhile, MySpace's biggest developers were beginning to turn their attention to Facebook as well. RockYou, for example, which had created a popular MySpace slide show application, immediately put much of its resources to building applications for Facebook instead. "Each social network is like another girlfriend," RockYou's founder explained. Although still the leading social networking site in the world,

MySpace was about to get dumped by its many suitors. "We're at a moment in time now where everything is shifting," BJ Fogg remembers thinking at F8. "In the future—in the next ten years—this will all seem so commonplace."

Fogg's brash prediction was wrong, however. It only took two years. By 2010 Facebook was practically ubiquitous, having established dominance in social networking and succeeding in laying its lens of "people you know and trust" over anything and everything on the Internet. "If the 2000's was the Google decade, then the 2010's will be the Facebook decade," media analyst Steve Rubel, Senior Vice President and Director of Insights for Edelman Digital, noted on his blog in February of that year. "I see it becoming the No. 1 website in the world in less than three years."

Rubel's brash prediction was also wrong. It only took ten months. According to data released by Hitwise, which provides online competitive intelligence, Facebook was the most-visited website in 2010, topping the list for the first time while accounting for 8.93 percent of all U.S. visits, ahead of such well-known online entities as Google, Yahoo, and YouTube. Facebook had doubled its share of visits from the previous year when it was the third most-visited site, and was up significantly from 2008, when it ranked ninth, trailing Google, eBay, then-social networking leader MySpace, and others.

Ironically, Facebook's rise was facilitated in part by another rival for Web supremacy: Google. In 2010, "Facebook" was the leading Google search term for the second year in a row; "Facebook login," was second, and "Facebook.com" and "www.facebook.com" also made it into the top 10 most popular searches. "Already, you can see the writing on the wall—pun intended," Steve Rubel wrote. "Look at the data."

In addition to leading in site visits, the data showed that Facebook had also surpassed Google as the top source for traffic to other major portals like Yahoo and MSN, according to Web measurement firm Compete Inc. It was also the leading source of links to many news media sites. By 2010, David Kirkpatrick's 2007 vision of knowing "which sites—or what news stories—people you trust found useful and which they disliked" had become an everyday reality. "Some experts say social media could become the Internet's next search engine," the Compete website reported. Rubel believed the shift was already happening, with Facebook leading the way.

"Google and search will remain important for years to come," he concluded. "However, what we're seeing is the beginning of big changes where social networking and Facebook will further disrupt advertising, media, one-to-one and one-to-many communications, not to mention search."

Rubel's enthusiasm was not misplaced. In an astonishingly short time period, Facebook had doubled in size and threatened to surpass Google in importance. Meanwhile its whiz-kid founder Mark Zuckerberg was making headlines almost daily, as the subject of best-selling books and a controversial if award-winning major motion picture, by giving away 100 million dollars to foster educational reform, being chosen as *Time* magazine's "Person of the Year," and then capping off an amazing year by seeing his still-privately held corporation attain a dizzying valuation of fifty billion dollars—five times its assessed worth just eighteen months before. As of January 2011 Facebook was deemed worth more than the likes of such long-established corporations as Time Warner or DuPont, despite showing negligible profit and sales of less than two billion dollars.

The "Facebook Way," as CNET's Caroline McCarthy dubbed it, may have appeared counter-intuitive to some but it was clearly working for Mark Zuckerberg. It meant maintaining "the feel of a small start-up, combating the potential for corporate sprawl and carefully constructing an environment that embraces minimalism," while staving off "an inevitable transformation into a corporate behemoth." People, not products, would be kept at the center of everything the company did. The most important person by far, of course, was the precocious Zuckerberg, who still controlled the company single handedly and who, despite his youth, remained steadfast in his strategy and decision-making.

"Two years ago, pundits were wringing their hands over Facebook, looking at the then-unprofitable company's finances and the tumultuous financial climate, with many suggesting that it was time for Zuckerberg to step aside in favor of a more seasoned CEO," McCarthy noted in January 2011. "But had Zuckerberg listened to any of these critics and stepped aside, the Facebook Way likely never would have taken root as a stable, post-start-up corporate culture, and the company quite likely couldn't have experienced the success that it can now boast.

"We didn't always want to admit it," she concluded, "But in 2010 the world accepted that Facebook—the company that introduced us all to such mundane pursuits as photo tagging, virtual farmsteads, and the voyeuristic tracking of the lives of people we only half-knew in high school—has changed the world."

5.

The Death of Privacy

As 2011 began, Mark Zuckerberg and the "Facebook Way" encountered critical new challenges. Despite his many successes—including the new valuation of the young company's worth set at an astonishing fifty billion dollars, an equally astonishing increase in membership from 350 to 600 million people, rising revenues and growing profit—Zuckerberg had made several missteps. The miscalculations raised fundamental questions about his reliability and ultimate intent and were serious enough to threaten permanent damage to relationships with users.

Maintaining the feel of a small start-up while managing an inevitable transformation into a corporate giant proved to be a tricky balancing act. So too did the goal of keeping people and not products "at the center of everything." Once-latent concerns over privacy, power, and profit now led regulatory agencies, ranging from the Federal Trade Commission (FTC) to the Securities and Exchange Commission (SEC), to scrutinize the company more closely. Was Facebook "on a trajectory to be pretty universal soon," as

Zuckerberg told his muse David Kirkpatrick? If so, would its ubiquity transform the platform into something more resembling a utility? Or are Zuckerberg and his now-powerful brand instead poised to stumble at the very brink of Web primacy?

Some critics believe that Zuckerberg and Facebook are indeed perched on a precipice and ready for a fall. Not surprisingly, the central issue revolves around trust. In one anti-Facebook screed on Wired.com, naysayer Ryan Sholin complained, "Facebook has gone rogue, drunk on founder Mark Zuckerberg's dreams of world domination." Where once it was "a place to share photos and thoughts," Sholin said, things began to go sour when "Facebook realized it owned the network," and with it the personal information that more than half a billion people had posted there.

Predictably, privacy and trust quickly became linked battlegrounds. Zuckerberg had repeatedly made sudden, sometimes ill conceived and often poorly communicated policy changes that resulted in once-private personal information becoming instantly and publicly accessible. The company's growing stature and importance only magnified the concerns. As Facebook profile pages began morphing more and more into overall online identities, the inherent tension between an individual's desire to protect personal information and the company's decision to make that information public came into sharp focus.

Facebook had changed its rules and privacy settings time and again, while Zuckerberg repeatedly pronounced privacy to be outmoded. He argued that we are living in a new era beyond privacy and that "the new social norm for the next generation is to share, freely and without regard for such antiquated concerns," as Mike Melanson posted on the *ReadWriteWeb* technology blog.

The controversy over privacy and trust had begun in November 2007 with Facebook's launch of a program called "Beacon" that used software to track users' activities on other websites. Zuckerberg arrogantly decided *not* to inform members that what they did on other sites would now flow back for all to see on Facebook, nor did he offer them the option to prevent that information from being shared.

An immediate backlash followed, which included the filing of a class action suit led by one man who had purchased a diamond ring for his wife as a surprise, only to be surprised himself when the purchase was then broadcast to hundreds of people throughout his Facebook network. Within weeks, the controversy over the use of tracking devices caused Zuckerberg to apologize and announce that he would allow users to opt out. Later he shut down the program completely. But the Beacon fiasco was just the beginning of the Facebook assault on privacy.

Controversy erupted anew, for example, in December 2009, when the company suddenly reneged on its assurances and made much of the previously private profile information public by default—including your name, your photo, the city that you live in, the names of your friends and the causes you'd signed onto. (Perhaps coincidentally, although conspiracy-minded critics saw it otherwise, the company's Director of Public Policy, Barry Schnitt, noted the same month that the time had come for Facebook to begin increasing its advertising revenue.) Following threats by advocacy groups such as the Electronic Privacy Information Center (EPIC) to file a complaint with the FTC, however, the decision to change Facebook's privacy policies was abruptly reversed.

Undaunted, Zuckerberg continued to defend his assault on users' privacy. In a January 2010 interview with Michael Arrington of the TechCrunch blog, he said that were he to create Facebook again, user information would be public by default, and not private as it had been for years before the recent and dramatic changes.

"When I got started in my dorm room at Harvard, the question a lot of people asked was 'Why would I want to put any information on the Internet at all? Why would I want to have a website?'" Zuckerberg recalled. "And then in the last 5 or 6 years, blogging has taken off in a huge way, and all these different services that have people sharing all this information.

People have really gotten comfortable not only sharing more information and different kinds, but more openly and with more people. That social norm is just something that has evolved over time. . . . We view it as our role in the system to constantly be innovating and be updating what our system is to reflect what the current social norms are."

Facing a backlash from Facebook members, Zuckerberg

remained resolute. "A lot of companies would be trapped by the conventions and their legacies of what they've built—doing a privacy change for 350 million users is not the kind of thing that a lot of companies would do," he concluded. "But we viewed that as a really important thing, to always keep a beginner's mind and what would we do if we were starting the company now and we decided that these would be the social norms now and we just went for it."

The new policy represented a significant shift from the way Zuckerberg had previously spoken about the importance of user privacy. In fact, since its inception, information on Facebook had only been visible to the people you accepted as friends. It was "fundamental to the DNA of the social network that hundreds of millions of people have joined over these past few years," one critic complained.

Zuckerberg once claimed that he believed privacy control was "the vector around which Facebook operates." Now he said Facebook was merely reflecting societal changes away from an emphasis on privacy. Even so, the manner in which the company was going about the changes left many wondering uncomfortably if Facebook's new philosophy about privacy had been created not because the "social norm" had shifted but for more crass and convenient reasons: commerce and control over the future of the web.

Despite the uproar hand-wringing, and recriminations, Facebook went even further in Spring 2010, and made more previously private information public. It was clearly no coincidence that the new round of changes also made it easier for advertisers to reach Facebook members in a variety of ways. First Facebook launched a product called "Instant Personalization," which automatically shared users' profile information with other sites, such as Pandora, Yelp, and Microsoft—without asking anyone's permission. (When one visited Pandora, for example, the music site could automatically access the music preferences listed on the "Likes and Interests" section of a Facebook profile and then recommend music selections it "thought" you might like with no action on your part.) Facebook also began requiring its members to join public groups based on the previously private interests they had included in their profiles. The company unintentionally exposed private chats of users as well. This time consumer protection groups, including the Electronic Privacy Information Center (EPIC) and fourteen others, did file an unfair-trade complaint with the FTC in response.

The complaint accused Facebook of unfair and deceptive trade practices that "violate user expectations, diminish user privacy, and contradict Facebook's own representations" and asked the FTC to order the company to "restore privacy settings that were previously available . . . and give users meaningful control over personal information." It said that Facebook's decisions to disclose previously restricted "personal information to the public" had violated users' expectations, diminished their privacy, and contradicted its own representations. It also urged the FTC to investigate Facebook's trade practices, to require the company to restore privacy settings that were previously available and to force it to "give users meaningful control over personal information."

Specific issues raised included the new Instant Personalization feature, the inability of users to make sections of their profile private, and Facebook's continued disclosure of information even after users chose to keep it private. Facebook also had made it difficult to block Instant Personalization. Even if a user opted out, for example, information could still be disclosed to third-party sites through that person's friends if *they* had not disabled the feature as well. The FTC complaint accused Facebook of concealing a "users' ability to fully disable Instant Personalization," and claimed its "privacy settings are designed to confuse users and to frustrate attempts to limit the public disclosure of personal information."

A timeline (See "Privacy Timeline" Sidebar) of Facebook's eroding privacy policy posted by Kurt Opsahl on the Electronic Frontier Foundation site detailed the remarkable transformation Facebook's position on privacy had undergone since its incorporation five years earlier. "When it started, it was a private space for communication with a group of your choice," Opsahl noted. "Soon, it transformed into a platform where much of your information is public by default. Today, it has become a platform where you have no choice but to make certain information public, and this public information may be shared by Facebook with its partner websites and used to target ads." To help illustrate his points, Opsahl highlighted excerpts from Facebook's privacy policies over the years "Watch closely as your privacy disappears, one small change at a time!" he warned.

Facebook executives responded aggressively, saying the many complaints came not from its members but from "media, organizations and officials." As Nicholas Carlson noted on the Business Insider website, "Facebook is getting lots of heat about privacy lately. But a company exec told Computerworld that any complaining you're hearing is not from real users."

Ethan Beard, director of Facebook's developer network, said, "The response from users speaks very, very loudly that they love what we're doing . . . to be honest, the user response has been overwhelmingly positive." He maintained that, "The reason that people use Facebook is to share information with their friends and to connect with things that are important to them . . . Sharing is not inherently a private activity." Business Insider's Carlson agreed, concluding, "What Ethan didn't say, but should have, is that one reason users aren't complaining about Facebook's recent privacy changes is that really, normal people don't really care about privacy any more."

The controversy over Facebook's privacy and trust issues, however, involved many "real users" and "normal people" and far more than just complaints from "media, organizations and officials." One protest started in Canada, quickly moved online and led to the creation of a website, www.quitfacebookday.com, where tens of thousands of users pledged to erase their profiles permanently from Facebook's database. "Why are we quitting?" their manifesto asked. "For us it comes down to two things: fair choices and best intentions. In our view, Facebook doesn't do a good job in either department. Facebook gives you choices about how to manage your data, but they aren't fair choices, and while the onus is on the individual to manage these choices, Facebook makes it damn difficult for the average user to understand or manage this. We also don't think Facebook has much respect for you or your data, especially in the context of the future."

The Quit Facebook website and similar user responses elsewhere made it clear that trust—and not simply privacy concerns—was at the root of the problem. "For a lot of people, quitting Facebook revolves around privacy," the manifesto noted. "This is a legitimate concern, but we also think the privacy issue is just the symptom of a larger set of issues. The cumulative effects of what Facebook does now will not play out well in the future, and we care deeply about the future of the web as an open, safe and human place. We just can't see Facebook's current direction being aligned with any positive future for the web, so we're leaving." Users could not, however, take their data with them when they left, as it remained Facebook policy to hold on to personal information even from deactivated accounts.

Users also expressed their unhappiness by creating new alternatives. One example was Diaspora, an open-source social network that pledged to allow users to control fully the information they shared by setting up their own personal servers, called "seeds." Raphael Sofaer, co-founder of Diaspora, said that centralized networks like Facebook were not necessary. "In our real lives, we talk to each other," he said. "We don't need to hand our messages to a hub."

Meanwhile, other Facebook users signaled they didn't believe privacy was "overrated" and began to modify their privacy settings on the network itself. Young people—Mark Zuckerberg's peer group of fellow Digital Natives, who supposedly didn't "really care about privacy any more"—led the way. Was the age of privacy truly over?

Not quite yet, apparently. Data released in July 2010 by researchers Eszter Hargittai and Danah Boyd demonstrated conclusively that rather than disregarding privacy, the young were increasingly switching their Facebook privacy settings in response to the changes. Their paper, "Facebook Privacy Settings: Who Cares?" examined "the attitudes and practices of a cohort of 18- and 19-year-olds surveyed in 2009 and again in 2010 about Facebook's privacy settings." Hargittai and Boyd concluded that, "Our results challenge widespread assumptions that youth do not care about and are not engaged with navigating privacy. We find that, while not universal, modifications to privacy settings have increased during a year in which Facebook's approach to privacy was hotly contested. We also find that both frequency and type of Facebook use as well as Internet skill are correlated with making modifications to privacy settings."

The research showed that most Facebook users—

youthful or not—modified their privacy settings regularly, with the practice becoming more common as time went on. "This suggests that either Facebook's changes to the site or the public discussion about them that took place between 2009 and 2010—or a combination of the two—may have influenced people's practices," Hargittai and Boyd said. In any event, they had uncovered an entire generation of youthful users who not only seemed to care deeply about their privacy but also were also actively taking measures to ensure it.

"If actions speak louder than words, it certainly doesn't look like the age of privacy has ended," Mike Melanson wrote in a blog post about the research findings. "Only that small percentage that didn't modify their privacy settings seems to be agreeing with the idea of broadcasting their information to the world. The rest, it would seem, still like to keep some things private."

Other assessments of the new policies were harsher. Writing in August 2010, Allan Badiner accused Facebook of betraying users and undermining their privacy for crass commercial reasons. "Facebook's motto is 'Making the world open and connected,'" Badiner noted on the progressive news website AlterNet. "But along with its policy of openness and potential for social change, Facebook has repeatedly come under fire for its lax policies toward the privacy of its members."

Badiner urged Facebook users to look "Behind the Wall" where, he warned, "users are creating a cumulative data repository of all the relationships in the entire world and the intimate details of everyone's lives—what FB calls the 'social graph.' The databases and algorithms employed at Facebook to store, crunch, and make inferences about you are far greater holders of data than any government agency."

Facebook "is both an infomediary and an intermediary," Badiner reasoned, occupying "a pivotal position as the preeminent hub in the new information economy." The danger, he said, is that it has also become "the primary custodian of more information than has ever before been collected about human beings. As intermediaries and hosts for our communications with lovers, family members, friends, and colleagues, social network providers have access to extremely sensitive information, including data gathered over time and from many different individuals." Facebook has a "fiduciary duty to its users," Badiner argued, which "has to come before the realization of Facebook's dreams for reengineering mobile communications and the web to become a more people-centric and integrated community."

One way of doing so might be to adopt a proposal by the Electronic Frontier Foundation to create a "Bill of Privacy Rights for Social Network Users," which would include "the right to be clearly informed about the options for privacy, what information is being shared to whom, and notified when any legal entity requests information about them." Such legislation would ensure that users retain control over the use and disclosure of their personal information, along with the right to have it removed from servers if they decide to leave a social network.

Is Privacy Overrated?

The jury was still out as to precisely how much "real users" opposed Facebook's new emphasis on openness or cared about Zuckerberg's commercial aspirations. David Kirkpatrick, author of the best-selling *The Facebook Effect*, suggested that for Zuckerberg, building a social networking empire is not about money—although it certainly *takes* money to do it. "Since connecting hundreds of millions of people with their social and business acquaintances online requires thousands of servers, other equipment, and eventually a large data center and a highly paid staff, and since charging people a fee to join Facebook would slow the growth of the network," Kirkpatrick wrote, "Facebook needs advertising revenues and so must operate as a business."

According to Kirkpatrick, Mark Zuckerberg conceives of Facebook as being about a merger of the private and public spheres. "Facebook is founded on a radical social premise," Kirkpatrick explains, "That an inevitable enveloping transparency will overtake modern life . . . Facebook is causing a mass resetting of the boundaries of personal intimacy."

If Kirkpatrick's analysis is correct, privacy may in fact be overrated, as observers as varied as Carlson and Lewis McCrary contend. In an August 2010 blog post entitled "The Metaphysics of Facebook: Privacy is Overrated," McCrary noted that, "Suddenly it seems so appropriate that Facebook was invented on a college campus. The more one reflects on it, the more the Facebook experience resembles what goes on in the hallways of college dormitories at universities everywhere: personal boundaries are reduced, many try on new slightly new personas every other week, and late-night bull sessions abound.

"Like Facebook, in college we all had a 'wall,' which enabled us to present ourselves to new 'friends'—mostly through cheap posters purchased the first week of classes," McCrary said. "We even had those little note boards on our doors where passers-by, even if only of casual acquaintance, could leave messages for all to see. Today, those non-digital forms of social networking all seem so 1999 . . . In this new world of social networking," he concluded, "We're all Madison Avenue marketers of our own public image."

Even if Zuckerberg et al are correct in their assessment that the private and public spheres are somehow merging, the relationship between the Facebook community and the Facebook brand will ultimately turn more on issues of trust than of privacy per se. "Whether less privacy is good or bad is another matter," Marshall Kirkpatrick posted on the ReadWriteWeb technology blog. "The change of the contract with users based on feigned concern for users' desires is offensive and makes any further moves by Facebook suspect."

Despite the backlash, Zuckerberg remained steadfast. In Fall 2011 he introduced still more new features to his platform, announcing at another developers' conference, "No activity is too big or too small to share. All your stories, all your life . . . This is going to make it easy to share orders of magnitude more things than before." Once again, however, Zuckerberg's attempts to make sharing easy went down hard with his users on Facebook, and another storm of controversy erupted, first over issues of change, control and communication and then over privacy and trust.

Once again the new changes seemed to limit user's input and control. In a post on the Salon website entitled "Facebook's enraging status update," Andrew Leonard limned reactions to the changes. "Like, oh, around 750 million other users of Facebook, I logged on to the world's biggest social media network this morning and was immediately annoyed," Leonard noted. "Facebook had changed its user interface, again." To disgruntled users like Leonard, "Facebook was indulging, again, in outright effrontery: employing its own secret algorithmic sauce to highlight what it considered the most important 'top stories.'"

The previous "Most Recent" button, which had allowed users to see what their friends have posted in chronological order, was now gone. In its place Zuckerberg had created a "Ticker" that provided a real-time stream of updates from friends. "Judging solely from comments from my friends, people don't want Facebook deciding what's most important," Leonard complained. "Facebook's suggestions were wrong, irrelevant and insulting, and why oh why oh why can't Facebook leave a good thing alone?"

To longtime Facebook aficionados, it was a case of déjà vu all over again: Zuckerberg was back to mandating major changes in surprising and disconcerting ways. "The pattern is set in stone," noted Leonard. "First there's a big uproar, then a flurry of suggested workarounds that will either revert the changes back to the idyllic past or otherwise nullify the most outrageous new abuses of our sensibilities Occasionally Facebook rolls back some particularly egregious privacy violation. But usually, the uproar soon subsides. We return to our gossip, snark and embarrassing family photos. And Facebook continues its inexorable growth The dynamic is beyond irritating: The fact that Facebook user complaints never amount to anything much probably emboldens Facebook in its behavior."

With their growing base of 800 million users, Facebook executives may well believe they are immune to such user dissatisfaction, owing to what Leonard aptly termed, "The golden fetters of the network effect." As he explained, "We're locked in by the comprehensiveness of the Facebook universe . . . And there's clearly more of the same (that is to say, constant discombobulating change) coming down the pike."

Shortly after the latest uproar had subsided, renewed concerns over privacy and trust began to shake the Facebook brand again. The latest privacy blunder centered on Facebook's belated admission that it was still tracking the web pages its members visited, even after they have logged out of the Facebook site.

As Daniel Bates reported for the *Daily Mail*, "The social networking giant says the huge privacy breach was simply a mistake—that software automatically downloaded to users' computers when they logged in to Facebook 'inadvertently'

continued on page 92

A Timeline of Facebook's Eroding Privacy Policy

In April 2010, Kurt Opsahl, a senior staff attorney with the Electronic Frontier Foundation, published the following on the organization's website: Since its incorporation just over five years ago, Facebook has undergone a remarkable transformation. When it started, it was a private space for communication with a group of your choice. Soon, it transformed into a platform where much of your information is public by default. Today, it has become a platform where you have no choice but to make certain information public, and this public information may be shared by Facebook with its partner websites and used to target ads.

Facebook Privacy Policy circa 2005: No personal information that you submit to Facebook will be available to any user of the Web Site who does not belong to at least one of the groups specified by you in your privacy settings.

Facebook Privacy Policy circa 2006: We understand you may not want everyone in the world to have the information you share on Facebook; that is why we give you control of your information. Our default privacy settings limit the information displayed in your profile to your school, your specified local area,

and other reasonable community limitations that we tell you about.

Facebook Privacy Policy circa 2007: Profile information you submit to Facebook will be available to users of Facebook who belong to at least one of the networks you allow to access the information through your privacy settings (e.g., school, geography, friends of friends). Your name, school name, and profile picture thumbnail will be available in search results across the Facebook network unless you alter your privacy settings.

Facebook Privacy Policy circa November 2009: Facebook is designed to make it easy for you to share your information with anyone you want. You decide how much information you feel comfortable sharing on Facebook and you control how it is distributed through your privacy settings. You should review the default privacy settings and change them if necessary to reflect your preferences. You should also consider your settings whenever you share information. Information set to "everyone" is publicly available information, may be accessed by everyone on the Internet (including people not logged into Facebook), is subject to indexing by third party search engines, may be associated with you outside of Facebook (such as when you visit other sites on the internet), and may be imported and exported by us and others without privacy limitations. The default privacy setting for certain types of information you post on Facebook is set to "everyone." You can review and change the default settings in your privacy settings.

Facebook Privacy Policy circa December 2009: Certain categories of information such as your name, profile photo, list of friends and pages you are a fan of, gender, geographic region, and networks you belong to are considered publicly available to everyone, including Facebook-enhanced applications, and therefore do not have privacy settings. You can, however, limit the ability of others to find this information through search using your search privacy settings.

Current Facebook Privacy Policy, as of April 2010: When you connect with an application or website it will have access to General Information about you. The term General Information includes your and your friends' names, profile pictures, gender, user IDs, connections, and any content shared using the Everyone privacy setting. . . . The default privacy setting for certain types of information you post on Facebook is set to "everyone." . . . [B]ecause it takes two to connect, your privacy settings only control who can see the connection on your profile page. If you are uncomfortable with the connection being publicly available, you should consider removing (or not making) the connection.

sent information to the company, whether or not they were logged in at the time. Most would assume that Facebook stops monitoring them after they leave its site, but technology bloggers discovered this was not the case."

Instead, the tracking information—worth billions of dollars to advertisers—was being sent back to the Facebook servers. Even after you were logged out, Facebook still tracked every page you visited. As Bates noted, "The admission is the latest in a series of privacy blunders from Facebook, which has a record of only correcting such matters when they are brought to light by other people."

Facebook members responded with a barrage of outrage on technology sites such as CNET, where one asked, "Who the hell do these people think they are? 'Trust us?' Why? Why should we trust a company that spies on us without our knowledge and consent?" Another added, "Holy wow . . . they've just leapt way past Google on the creepy meter."

Like all brands, Facebook is built on trust. As its executives struggled to explain the latest "inadvertent" privacy row over its "creepy" web-tracking practices, that trust was shaken once again "by criticism and speculation regarding how it uses browser cookies to get data about users," as Josh Constine posted on Insidefacebook.com. "A lack of thorough documentation explaining what each of its cookies does has led some observers to assume that the company is tracking offsite browsing behavior in order to target ads. Facebook needs to provide explanations for both the average user and privacy researchers about how exactly its cookies work in order to prevent these press flare-ups from giving users a negative impression and bringing on regulatory scrutiny from governments."

In an early October 2011 post titled "Brutal Dishonesty," Michael Arrington's new "Uncrunched" blog reported that Facebook had filed a patent application involving a method for "tracking information about the activities of users of a social networking system while on another domain."

The language seemed to indicate that the information at least had the potential to be used to target Facebook advertising.

Although many believed the "cookies" left by Facebook and third-party sites that track users' web browsing behavior violated their privacy by supplying data to target ads, Facebook continued to claim they were only used to enhance site security. As Constine noted, however, "Unfortunately for Facebook, the claims are still giving off a negative impression of the service and sparking complaint letters to government agencies from privacy advocate groups. A patent application for the company's social plugins that included language about tracking and targeting ads has also helped fuel the controversy."

In response to the many critical comments, Facebook engineer Gregg Stefancik admitted, "We haven't done as good a job as we could have to explain our cookie practices." Once again, Facebook executives were forced to apologize and backtrack, when the brouhaha could easily have been avoided simply by being more transparent and less arrogant.

Constine concluded, "Facebook wants to be seen as above the controversies surrounding the industry—and because so many users opt in to share their data to Facebook, the company is a cut above in many ways. Yet the combination of unclear explanations, past issues, and the patent are getting in the way of its effort to explain its case. . . . The onus is now on Facebook to fully explain how it does and does not track users across the web and use that information back on Facebook—and prove what it says through the technology that it deploys across the web."

As the battle over privacy and trust continued, some observers said they supported Zuckerberg's notion that privacy was dead. An advertising consultant named Cindy Gallop was among them, telling a BBC audience that, "At a time when there are many debates about the privacy settings on Facebook and the ethics of Wikileaks, I actually come from the complete opposite end of that argument."

To Gallop, "The new reality that all of us live in today, personally and professionally, is one of complete transparency. Everything we do and say today, whether we are a person or a brand, business or company, is potentially in the public domain courtesy of the internet." She explained that her stance offered "an overall benefit which I call action branding. Personal action branding is for individuals and corporate action branding is for brands and businesses, and it is the advertising of the future. . . . It is not about telling, it is about being. Brands will be judged just as people are so action branding is communication through demonstration."

Still, the overall reaction to the latest Facebook privacy controversy remained negative. The high-handed manner in which members' personal information had been treated, the lack of consultation or even communication with them beforehand, Facebook's growing domination of the entire social networking sphere, Zuckerberg's constant and very public declarations of the death of privacy and his seeming imposition of new social norms all feed growing fears that he and Facebook itself *simply can not be trusted*. As Zuckerberg's fellow CEOs from the legacy media had already learned, losing the trust of your audience is the first step in losing your audience itself—and eventually the power of your brand.

Zuckerberg disingenuously maintains that he is not changing privacy mores but simply responding to changes in them, while at the same time asserting that users' complaints about control of their personal information are old-fashioned. But as Ryan Sholin pointed out in *Wired*, "Setting up a decent system for controlling your privacy on a web service shouldn't be hard. And if multiple blogs are writing posts explaining how to use your privacy system, you can take that as a sign you aren't treating your users with respect. It means you are coercing them into choices they don't want using design principles. That's creepy.

"Facebook isn't about respect," Sholin continued. "It's about re-configuring the world's notion of what's public and private." He concluded by denouncing it as a "rogue company" that "got to be the world's platform for identity by promising you privacy and then later ripping it out from under you."

Rogue or not, Facebook's meteoric growth has yet to be slowed by the continuing ruckus over privacy—even as reports such as a recent one from the American Customer Satisfaction Index rank the company in the bottom five percent of social media sites. The survey showed users' concerns centered around privacy, interface changes, navigation problems, and aggressive advertising. But Mark Zuckerberg doesn't seem overly concerned. Although the focus of intense scrutiny, he says he doesn't read much of the press, posts or books about Facebook, and doesn't care about negative portrayals of him (such as that in the hit film *The Social Network*.) Ironically, as Allan Badiner observed, "To the great modern prophet of staying connected, being disconnected sometimes is a good thing."

"Over time," Zuckerberg says, "people will remember us for what we build and how useful it is to them." Considering just the relatively low number of Facebook defectors versus the onrush of new users, his confidence may not be misplaced. Instead of quitting Facebook, the vast majority of its members still holds out hope the company will demonstrate a greater future respect for their privacy, even as it continues to personalize the web, which intrinsically requires obtaining information from users. The terms of the unstated deal we all make with social media—exchanging our personal information for a free platform to share it on—is in a constant state of flux. Facebook and similar networks are collecting more information from and about us than ever before, and then acting as if they own it. If we choose to share data on a network with our friends and followers, must we necessarily cease to control it? Imagine how much we might share if we also shared in the revenue generated. . . .

Trust is indeed "the new black," as Craig Newmark has noted often. "People use social networking tools to figure out who they can trust and rely on for decision making," he says. "By the end of this decade, power and influence will shift largely to those people with the best reputations and trust networks, from people with money and nominal power."

Trust is also essential for the success of any brand—including that of Facebook. Mark Zuckerberg may think that the recent privacy flaps haven't affected the network much, but they actually represent a huge potential threat to what he has built. As Newmark warns, "We are already seeing a shift in power and influence, a big wave whose significance we'll see by the end of this decade. Right now, it's like the moment before a tsunami, where the water is drawn away from the shore, when it's time to get ahead of that curve."

Augie Ray, senior analyst at Forrester Research, calls what has been happening at Facebook "death by a thousand privacy cuts." So far, however, the bleeding hasn't been too bad. In fact, Facebook is said to be working on plans to celebrate its one-billionth member, whom it expects to log on by the end of 2011. Should a viable alternative emerge, its brand could implode almost overnight and Facebook could just as easily become the new MySpace, celebrated as recently as 2009 as "the most popular website in America" but practically moribund just two years later.

What will happen if, as *Wired*'s Ryan Sholin suggests, "the best of the tech community" does find a way "to let people control what and how they'd like to share?" Such a response would represent a clear challenge to Facebook's proprietary, profit-making protocols. "Think of being able . . . to build a profile page in the style of your liking," Sholin posited. "You'd get to control what unknown people get to see, while the people you befriend see a different, more intimate page. They could be using a free service that's ad-supported, which could be offered by Yahoo, Google, Microsoft, a bevy of startups or web-hosting services."

As even critics as harsh as Sholin admit, such an event may be unlikely, "nor would it be easy for that loose coupling of various online services to compete with Facebook." Moreover, as he concedes, "Facebook has taught us some lessons. We want easier ways to share photos, links and short updates with friends, family, co-workers and even, sometimes, the world. But that doesn't mean the company has earned the right to own and define our identities."

In seizing rather than earning that right, however, Facebook risks losing everything—its credibility, the trust of its users, and ultimately its brand. Even while it battles with Google and others for Web dominance, Facebook is simultaneously set to stumble over the centrality of trust and reliability. Unless altered, Mark Zuckerberg's blind ambition and inability to listen to *his* many friends and followers could yet lead not to dominance but to downfall.

6.

The YouTube Effect: 48 Hours Every Minute

In a decade that saw social media move rapidly from the fringes of the Internet to the mainstream of online activity, YouTube—and not Facebook, with its eight hundred users—was the innovation that touched the most lives. In remarkably short order, the social video sharing platform also became a driving force for change all over the world.

Before YouTube existed, it was difficult for those lacking tech savvy to post videos online. Developed in early 2005, YouTube's simple user interface made it easy for anyone with an Internet connection to upload video content for consumption and sharing with a worldwide audience that could comment on and rate each video. YouTube turned video sharing into an important part of global culture and communications; in turn, enthusiastic users made the site a cultural and political force to be reckoned with.

Although often blocked by governments upset at videos of protests and other supposedly "offensive materials" posted there, the site was never offline for long. By 2008 it was widely hailed as an online "Speakers' Corner" and called "an ever-expanding archive-cum-bulletin board that both embodies and promotes democracy" when honored with a Peabody Award, the world's oldest and most prestigious electronic media prize.

The service first became popular when members at MySpace, then the leading online social network, discovered they could embed YouTube videos into their profiles. But MySpace executives viewed YouTube as a competitive threat to their own video service and soon banned it from user profiles. Although protests led to a quick reversal of the policy, it was already too late for MySpace to recover from its colossal blunder. YouTube launched its own site in November 2005 and soon surpassed MySpace to become the world's fifth most popular website. Less than a year later, in October 2006, Google purchased YouTube for $1.65 billion—a seemingly huge figure given the site's brief existence and relative lack of revenue.

In retrospect, the Google purchase seems quite visionary, however. YouTube now ranks third in total traffic among all websites, behind only Facebook and Google itself. It is by far the dominant provider of video on the Internet, serving 6.6 times more people than its leading competitor, Hulu.com, according to the Nielsen rating service. The site is still growing at an incredible rate, with the number of uploads doubling in two years. By its sixth birthday in May 2011, users from all over the world were posting more than 48 hours of video on YouTube servers every minute—

up from six hours per minute in 2007, fifteen hours by 2008, and twenty-four hours as of May 2010.

By November YouTube's director of product management, Hunter Walk, announced that the number of hours uploaded every minute had risen from twenty-four to thirty-five hours. 2,100 hours were uploaded every 60 minutes, or an astonishing 50,400 hours of fresh video every day. "If we were to measure that in movie terms (assuming the average Hollywood film is around 120 minutes long), 35 hours a minute is the equivalent of over 176,000 full-length Hollywood releases every week," the company's blog explained. "Another way to think about it is: if three of the major U.S. networks were broadcasting 24 hours a day, 7 days a week, 365 days a year for the last 60 years, they still wouldn't have broadcast as much content as is uploaded to YouTube every 30 days."

Walk challenged his users to go further. "Clearly, you are able to tackle some of our most daunting challenges. So it is with that in mind that we throw down another heavy gauntlet: upload 48 hours of video every minute," he blogged. "That's right: two full days and 100% growth of what we achieved back in March of 2010." The goal was easily met—again, within just six months.

YouTube executives should be careful what they wish for, however. Although such phenomenal growth in usage is one measure of success, it may also paradoxically augur the platform's demise. As an ever-expanding archive, YouTube has become a leader in illustrating the difficulty of separating signal from noise in the new social media world. What better example of an unfiltered flood of information can be found than a site that shares such a prodigious amount of unvetted, unverified and too-often untrue information? How can anyone possibly determine which pieces of the voluminous information and news buried within the forty-eight hours of video posted there in the minute it took to read this paragraph are actually true? How can we ever hope to find credible news and information we can rely on? Can we trust anything found on YouTube?

The Macaca Moment

It wasn't like this in the beginning. Steve Grove, YouTube's director of News & Politics, joined the company early on. A slender, affable and articulate thirty-something from Northfield, Minnesota, Grove had been an All-American athlete at Claremont McKenna College, where he captained the track & field team. Following a short stint as a correspondent at the *Boston Globe*, he received a Master's degree in Public Policy from Harvard University's Kennedy School of Government in 2006. Upon graduation, he and two classmates persuaded school officials to commission them to travel the world, shooting and editing video profiles of other recent graduates.

"I discovered YouTube because we wanted a way to document our trip," Grove recalls. "We had just found out about this new way to upload and share video via the Internet. So we started posting video to YouTube from the road, from Internet cafes in Vietnam, Serbia, and India. I became enamored by the phenomenon of this new site that let anyone become a global broadcaster." Upon his return to the United States, Grove wrote to YouTube executives. "I asked if they needed someone to work on politics," he says. "This was right around the time of George Allen's infamous Macaca Moment, so YouTube was gaining prominence as a political platform."

The "Macaca Moment" Grove refers to involved a campaign rally in Virginia in August 2006. Republican Senator George Allen was running for re-election and expected to defeat James Webb, his Democratic challenger, rather handily. Allen was also planning a presidential bid for 2008. But when he used an ethnic slur in referring to S.R. Sidarth, a Webb campaign volunteer who had been assigned to trail Allen with a video camera and document his travels and speeches, the senator's political future hit the skids.

On August 11, 2006, at a campaign stop in Breaks, Virginia, Allen began a speech by saying he was "going to run this campaign on positive, constructive ideas." Then he pointed at Sidarth in the crowd. "This fellow here, over here with the yellow shirt, Macaca, or whatever his name is. He's with my opponent. He's following us around everywhere. And it's just great," Allen said, as his supporters began to laugh. "Let's give a welcome to Macaca, here. Welcome to America and the real world of Virginia."

In some cultures, the word "macaca," which is the name of a type of monkey, is considered a racial slur against immigrants and Africans. As Webb's communications director Kristian Denny Todd noted, Sidarth was neither an African nor an immigrant, but an Indian-American who had been born and raised in Fairfax County, Virginia. "The kid has a name," Todd told reporters. "This is trying to demean him, to minimize him as a person." Todd added that the use of "macaca," whatever it meant, and the reference welcoming Sidarth to America, were intended to make him uncomfortable.

When Sidarth's video of the incident was posted on YouTube, it promptly caused a sensation. Soon legacy media outlets picked up the story and re-distributed the video, leading to another spate of negative publicity. Allen quickly apologized, although he said the word "macaca" had no derogatory meaning for him. Asked to define it, he replied simply, "I don't know what it means."

Nevertheless, the damage was already done. Because of YouTube's reach, tens of millions of people who hadn't attended the event learned what the word meant. Although Allen's apology had come within days, it was still far too late; the video had already been shared all over the world. The controversy helped turn what most pundits had seen as Allen's inevitable re-election into a surprising victory for his opponent. In the process, it also scuttled Allen's looming presidential candidacy and made YouTube a household name and a politico-cultural force to be reckoned with.

A few months later, Grove was hired as YouTube's head of News and Politics, charged with directing the news and political strategy and programming for the company. He arrived just as interest in the site began to peak among experienced political operatives and legacy news media executives, who had been shocked at the speed at which George Allen's career arc had been altered by the new Web phenomenon.

The News Ecosystem

As America headed into an historic presidential campaign, Grove found himself at the white-hot nexus of media and politics. He led YouTube into content partnerships with a wide variety of media, political, and governmental organizations, aiming to develop new platforms and programming that would increase citizen engagement in politics, government, and news reporting. Numerous mainstream media outlets clamored to partner with YouTube, including broadcast networks like CBS and the BBC, wire services like the Associated Press, and national newspapers such as the *New York Times*, *Washington Post*, and *Wall Street Journal*. The media uploaded their branded content to YouTube because its "news ecosystem," as Grove termed it, offered them a new, interactive, and social means of engaging with an audience of news consumers—especially younger ones, many of whom had already abandoned the legacy news operations.

"Each day on YouTube hundreds of millions of videos are viewed, at the same time that television viewership is decreasing in many markets. It's where eyeballs are going," Grove explains. "If a mainstream news organization wants its political reporting seen, YouTube offers visibility without a cost. The ones that have been doing this for a while rely on a strategy of building audiences on YouTube and then trying to drive viewers back to their websites for a deeper dive into the content. And these organizations can earn revenue as well by running ads against their video content on YouTube."

For the legacy news media, audience engagement was much easier to achieve by using the YouTube platform. Its open API (application programming interface), called "YouTube Everywhere," allowed them to integrate its upload functionality into their own online efforts. "It's like having a mini YouTube on your website," says Grove.

During the 2008 presidential election cycle, Grove forged a high profile relationship with CNN, centering around two debates during the primary season. Citizens were able to upload video questions for the candidates, which were vetted by YouTube and CNN and played during the debate. YouTube also hit the U.S. campaign trail along with Hearst affiliate WMUR-TV in New Hampshire, soliciting videos from voters during the first-in-the-nation primary election there. "Hundreds of videos flooded in from across the state," Grove recalls. "The best were broadcast on that TV station, which highlighted this symbiotic relationship: on the Web, online video bubbles the more interesting content to the top and then TV amplifies it on a new scale."

Grove later made similar arrangements with news organizations in Iowa and Pennsylvania during those state primaries, as well as during the multi-state primary known as

"Super Tuesday," when numerous news organizations leveraged the power of YouTube and its "voter-generated content." He created partnerships with other local, national and even international news media organizations, including with the BBC around the mayoral election in London and with a large public broadcaster in Spain during that country's presidential election.

The YouTube Effect

In addition to forming powerful media partnerships, Grove also helped pioneer the inclusion via YouTube of another new phenomenon called "citizen journalism." With media production tools widely available for the first time, non-professionals seized on YouTube's global distribution power to upload videos that in some instances had great impact on campaigns—in large part because they were then reported on by the mainstream media. This phenomenon became known, thanks to the *Washington Post*, as the "YouTube Effect."

"Not long ago, an anonymous video on the Internet would have elicited little more than amusement from the candidate under attack," political reporters Chris Cillizza and Dan Balz noted in the *Post* on January 22, 2007. "But the 2006 midterm campaign—in which then-Senator George Allen saw his hopes for reelection, not to mention the White House, torpedoed by his now-infamous 'macaca' moment captured on a widely seen video—changed the rules. The already-underway 2008 presidential campaign is likely to be remembered as the point where Web video

became central to the communications strategy of every serious presidential candidate. . . . Call it the YouTube effect, and it is only growing."

In addition to Allen's Macaca moment, Cillizza and Balz cited other examples of the Effect, from a pointed exchange between Associated Press reporter Glen Johnson and presidential candidate Mitt Romney to a spoof of a famous Apple Super Bowl Ad, which compared Senator Hillary Clinton to the oppressive system described in George Orwell's "1984."

In less than two years YouTube's video-sharing platform had "revolutionized the transfer of information via video, spawned a number of imitators and forced candidates to recalibrate choices, from their announcement strategies to their staffing decisions," Cillizza and Balz concluded. "Playing defense is only one use of Web video. Equally important, the candidates and their staffs see Web-based video as an inexpensive and potentially significant tool for telling their campaign story without the filters of the traditional media."

The new reliance on Web videos to tell unfiltered stories of candidates for office raised immediate issues of trust and credibility. As the presidential election moved into high gear, fact-checking the veracity of claims made in those videos and other campaign communications became a priority for the media that campaigns were now intent on bypassing. News organizations as varied as the *St. Petersburg Times*, *Congressional Quarterly* and the *Washington Post*, (which began awarding "Pinocchios" to candidates who bent the truth) started websites that examined third-party and campaign ads circulated via YouTube, while a "Veracifier" page created by the *Talking Points Memo* website became one of the most-trafficked pages within Grove's own News & Politics section.

By the middle of 2008 YouTube was firmly established as a major player in the worlds of both media and politics, simultaneously collaborating *and* competing with legacy firms. Grove launched a Citizen News channel to highlight user-generated content, and announced a partnership with the Pulitzer Center for a journalism contest encouraging YouTube users around the globe to submit videos telling the story of a person or cause untold by the mainstream press. Within a week, more than 200,000 people had watched a YouTube video detailing the contest.

The Summer 2008 issue of *Nieman Reports*, the nation's oldest magazine devoted to a critical examination of the practice of journalism, featured dozens of articles and essays in a section dedicated to the subject of "Politics and the New Media." In one piece, headlined "Campaign 2008: It's on YouTube," Albert L. May remarked on that "brief, heated exchange" between AP political reporter Glen Johnson and presidential candidate Mitt Romney, which had been noted earlier by Cillizza and Balz in the *Washington Post*. May said that while the dustup was "familiar to any veteran political reporter who has spent time inside the sometimes fractious 'bubble' of presidential campaigns," something was different this time.

"The scene was uploaded onto YouTube and an incident that would have been relegated to a campaign footnote took on a life of its own," said May. "In the new camera-rich environment, the video could have just as easily come from a voter's cell phone or the hand-held device of the ever-present campaign trackers." The bubble in May's analysis had suddenly "become a fishbowl."

After airing on cable news talk shows, the Johnson-Romney video went viral, and Johnson became "a minor YouTube celebrity and a member of the million plus hits club on the Google search engine," he noted. In the old media world, with its lack of "a centrally organizing website with easy posting of video by anyone with a digital camcorder, laptop and inexpensive software, viewing didn't happen among those who lacked a connection to the political underground," May concluded. "All of this changed with the February 15, 2005 launch of YouTube."

Steve Grove also contributed an essay to the same issue of *Nieman Reports*. In it, he warned, "With YouTube's global reach and ease of use, it's changing the way that politics—and its coverage—is happening." Grove pointed out that each of the sixteen candidates for president had created YouTube channels; seven had even announced their candidacies on YouTube, and their staffs had uploaded thousands of videos that were viewed tens of millions of times.

"The most exciting aspect is that ordinary people continue to use YouTube to distribute their own political content," he added. "These range from 'gotcha' videos they've taken at campaign rallies to questions for the candidates, from homemade political commercials to video mash-ups of mainstream media coverage." What it all meant was that "average citizens are able to fuel a new meritocracy for political coverage, one unburdened by the gate keeping 'middleman,'" Grove said. "Another way of putting it is that YouTube is now the world's largest town hall for political discussion, where voters connect with candidates—and the news media—in ways that were never before possible."

Poised at the emerging frontier of media and politics, new and old, YouTube still faced a huge unresolved issue. The YouTube Effect, or what Albert L. May dubbed the "YouTubification" of politics, had led to a great increase in what he called "unmediated information" flowing to an ever-larger number of Americans. On one hand, in enabling the widespread availability of videos from and about campaigns, YouTube had "revitalized and enriched the 'truth testing' of candidate advertisements and other messages," May concluded. "Now news organizations, from newspapers to local television, use the Web to extend the reach of their traditional coverage of candidate truthfulness." On the other hand, while some might laud the "democratization they see happening with the rise of 'citizen journalists' who make their own 'mash-up' videos," he noted, "It is often difficult to ascertain the sources and to figure out just what has been mashed up."

Steve Grove concluded his essay with a section called "Trusting What We See," wherein he explained that both citizens and professional journalists played an important role in determining the trustworthiness of news content on YouTube. "People tend to know what they're getting on YouTube, since content is clearly labeled by username as to where it originated," Grove wrote. "Users also know that YouTube is an open platform and that no one verifies the truth of content better than the consumer. The wisdom of the crowd on YouTube is far more likely to pick apart a shoddy piece of 'journalism' than it is to elevate something that is simply untrue.

"In fact, because video is ubiquitous and so much more revealing and compelling than text, YouTube can provide a critical fact-checking platform in today's media environment," he asserted. "And in some ways, it offers a backstop for accuracy since a journalist can't afford to get the story wrong; if they do, it's likely that someone else who was there got it right—and posted it to YouTube."

But trust is a two-way street, and "scrutiny cuts both ways," Grove said. "Journalists are needed today for the work they do as much as they ever have been. While the wisdom of crowds might provide a new form of fact checking, and the ubiquity of technology might provide a more robust view of the news, citizens desperately need the Fourth Estate to provide depth, context and analysis.

"YouTube has become a major force in this new media environment by offering new opportunities and new challenges," he concluded. "For those who have embraced them—and their numbers grow rapidly every day—the opportunity to influence the discussion is great. For those who haven't, they ignore the opportunity at their own peril."

Harnessing the Beast

When Grove wrote those words, YouTube users only posted ten hours a video every minute. By the end of 2010, that figure had more than tripled, and his primary challenge had become "how to harness the Beast of YouTube to news." With content pouring onto the site, Grove thought constantly about how to make it more useful, "how to make sense from chaos and profusion." What filters would bring trustworthy, high-quality news to the fore? How helpful were the "recommender systems," machines that could learn from your behavior and then employ algorithms to intuit what you seek, based on clues drawn from location, history, personal interests and the like? How much of a role did other social networks play? Grove thought Facebook and Twitter were "tremendous news platforms which help you find what friends find useful," but looking forward, was there some way to move beyond social filters? He wondered how he could somehow find a way to "leverage curation to create or empower tastemakers who have gained reputation on Facebook and Twitter." If he succeeded, he believed they might in turn grow into micro brands themselves one day.

"Right now at YouTube we're working more on incentivizing curation," Grove told me in a recent interview. "We do a pretty good job of incentivizing uploading and watching of content, but the site doesn't leverage taste-making as well as it could. The YouTube channel, for example, shouldn't just be a place to host your own content, but a place to essentially create your own television network: programming content you like for the world to see. The better job you do as a programmer, the more viewers you'll get."

Will the YouTube audience itself help Grove figure out what matters? Can YouTube help by creating better tools, such as user-generated news feeds, to assist in the discovery of useful news content? "Our engineers are constantly working on the discovery challenge," he said. "We're always developing new feeds and tools to improve discovery and search on the site, as well as building destination pages to get a quick snapshot of what's rising, such as YouTube Trends, YouTube.com/news, and our Citizentube experiment."

Citizentube was radically transformed in June 2010 when Grove announced a partnership with University of California at Berkeley's Graduate School of Journalism to it into a citizen journalism breaking news feed. Although YouTube had drawn attention to amateur journalism before, most notably during the 2009 Iranian election protests, Grove aimed to take the concept higher by actively promoting timely user-generated video journalism. The feed provided a steady stream of breaking news videos, with a focus on non-traditional sources and the very latest uploads.

Meanwhile he continued to experiment, explore and create partnerships, all in an effort to tame the YouTube beast. "Our users and partners will continue to create their own filters of content from the site that they find useful," he says. "We'll continue to build tools to make that easier. And we'll we continue to look for ways to harness the curatorial power of the web to create better and more dynamic feeds of contents for users to discover."

By Fall 2011, YouTube had begun finalizing contracts for the launch of the first of new channels featuring regularly scheduled content on broad themes. The platform planned to pay outside content creators to create and curate videos for the channels, complete with actual schedules of programming, as it tried to build a comprehensive video service and reposition itself as direct competitor to broadcast and cable television.

Not surprisingly, a new Politics Channel was first out of the box, offering viewers what the Business Insider website described as "a one-stop place to view ads, speeches, gaffes and anything else in the political sphere." As Grove's colleague Ramya Raghavan, YouTube's news and politics manager, wrote in a blog post, "From the infamous 'Macaca' moment of 2006 to the recent Fox News/Google Debate, YouTube is a place where you can keep track of the latest political stories and connect with the candidates."

With YouTube now the Web's second most-used search engine, trailing only the mighty Google, the question of how to filter and manage the flood of news and information found there looms larger than ever. It's especially pressing since, like its parent company Google, YouTube does nothing to verify what appears on its site; instead it offers an open platform for anyone to post anything they like, whether fraudulent or not. "YouTube is a platform built on the principals of free expression—we will never be the arbiter of which content is true or false, good or bad," Grove maintains.

Yet he remains optimistic. "Technology will continue to do a lot to bring order to the chaos of information on the web, he says. "Aggregating the news judgment of users and of news professionals is one way to cut through diverse amounts of information and create credible streams of content. But there are many other ways, too . . . The world is too complex, and our sources of information too diverse and varied, for there to be just one stream of credible information developed by algorithms and aggregation."

In the meantime, the flood of unmediated information posted on YouTube continues unabated. At the same time, a new and even more social media-based American presidential campaign is shifting into gear. Even George Allen has re-entered the fray, announcing in early 2011—via email and Web video, ironically—that he intends to get his old job back. Although Allen made no mention of "macaca," the YouTube Effect still looms large. "The central question," the *New York Times* noted, "will be how much the 'macaca' matter weighs on Mr. Allen's reputation in the minds of Virginia voters."

7.

Twitter: News No Longer Breaks, It Tweets

On January 14, 2009, the first report of the miraculous res-
cue of 155 passengers from a US Airways jet floating in the
Hudson River provided the ultimate evidence—if indeed
any was still needed by then—that emerging social media
were not only supplementing but in some cases actually
supplanting legacy media in both reporting and distributing news. Twitter, the short messaging service or "micro-
blogging" platform launched for public use just two and a
half years earlier, had beaten the rest of the world's media to
the sensational story that an airplane had gone down in the
water shortly after takeoff. Despite the fact that an international wire service, a leading national newspaper and the
news divisions of several broadcast networks all had their
worldwide headquarters nearby—in fact almost literally opposite the crash site—a Twitter user named Janis Krums was
first on the scene. Krums quickly tweeted news of the crash
("There's a plane in the Hudson. I'm on a ferry going to pick
up the people. Crazy") and posted a photo of its passengers
huddled on a wing just moments after the aircraft plunged

the river—and well before anyone from the legacy news media arrived on the scene.

Krums, who was a passenger on a nearby ferry, took a picture with his iPhone and posted it instantly on the photo-sharing TwitPic.com site before getting off the boat to help the plane's passengers reach safety. As he did so, thousands of followers replied to Krums' Twitter account to congratulate him on his scoop and thousands more—including representatives of the legacy media he had beaten to the story—created links to the remarkable image. The photo spread around the social media world so rapidly—tens of thousands of people in the next four hours—that the heavy traffic soon crashed the site.

When it comes to breaking news—from heroism on the Hudson to calamity in California and from terror in Mumbai to protest in Tunis—Twitter now often leads the pack. Early adopters of the service learned about the service's news utility soon after its debut, but it took the "Miracle on the Hudson" to bring it fully to the attention of the rest of the world and give new meaning to the young company's brash claim that "Twitter is the best way to discover what's new in your world."

What many had previously dismissed as an insignificant messaging service had suddenly morphed into one of the most important mass communications systems in the world—and to the surprise of its creators, was transformed into a leading source of breaking news. Now simultaneously collaborating *and* competing with legacy media, Twitter is at the edge of the blurry frontiers separating news and entertainment, professionals and amateurs, and, perhaps most importantly, opinion and fact.

If you're still unclear about Twitter's phenomenal growth, importance and meaning, here's some background: the free social networking service enables anyone to post short messages known as tweets, 140 characters in length, to groups of self-designated followers. Tweets can be sent from and received by many different digital devices, ranging from desktop and laptop computers to smartphones and just plain cell phones. Tweeting is a much like instant or text messaging, but instead of one-to-one communication, it is one-to-many.

Twitter started in 2006 as a side project of Odeo, a podcasting site co-founded by Evan Williams. Williams had made an earlier fortune by creating Blogger, one of the first and most facile blogging tools, and then selling it to Google. Although Odeo had received millions of investment dollars, Williams was admittedly unexcited by its involvement in podcasts and asked everyone working there for new ideas. One day, while sitting on a children's slide at a park eating Mexican food, an engineer named Jack Dorsey showed his colleagues at Odeo a cool new way to use text messages to send status updates.

Since then Twitter has grown to become what the *New York Times* termed "one of the rare but fabled Web companies with a growth rate that resembles the shape of a hockey stick." The number of people signing up to use it increased exponentially; in 2009, for example, Twitter ballooned from 5 million to 71 million registered users. By September of the next year, it claimed 145 million users, and two months later another 30 million people had registered—a 20 percent increase. Executives estimated the service was adding more than 300,000 new users every day. By 2011 there were more than 200 million registered accounts—half of them so-called active users, many of whom log in daily.

Twitter.com soon became the third most trafficked social networking site in the world, surpassing MySpace with nearly 96 million unique visitors, according to data from comScore Inc. Both still trailed Facebook, which passed MySpace in early 2008 to take the top spot among social media and grew to 598 million unique visitors by August 2010. (Microsoft's Windows Live Profile, which integrates with the company's web-based email and other services, was second with more than 140 million visitors.)

As of October 2011, Twitter's 100 million active users were sending more than 200 million messages every day. These tweets form the basis of Twitter's real-time information network. Each can also connect to deeper context and embedded media such as photos or videos. But even if you don't tweet, you can still access many other voices freely offering news and information about a wide range of topics of interest.

Based in San Francisco, Twitter is now used by people in nearly every country in the world, and comes in seventeen language versions, including English, French, German,

Italian, Japanese, Spanish, Chinese, and Hindi. Seventy percent of its traffic comes from outside the United States. With its astonishing user growth, Twitter is poised to join Google and Facebook as one of the Internet's next hugely important independent companies. Although its annual revenue has yet to exceed 100 million dollars, it is projected to surpass one billion dollars by 2016. The company has already become so popular and ubiquitous that it is valued at between eight and ten billion dollars.

Despite its success to date, Twitter is still nowhere near fulfilling its true potential. Although its valuation is certainly not justifiable based on revenues, investors highly value its social services and data about users—and there is much, much more to come. Of people who use the Internet, nearly one of five use Twitter, according to a recent survey by Pew Research Center. Facebook, on the other hand, is used by 96 percent of all Americans. These figures make Twitter's valuation sensible, as Felix Salmon pointed out on his Reuters blog, noting, "If Twitter is 20% the size of Facebook, and Facebook is worth $50 billion, then Twitter can be worth $10 billion, no?"

Any way you slice it, clearly Twitter is quickly becoming central to how people communicate—it's a "key part of the new social architecture," as Salmon says. "Twitter serves a very important purpose in the lives of the people who have adopted it, and it's likely to serve the same purpose for ever more people as its user base grows and people start feeling left out if they're not on it . . . priority number one for the company is to become an indispensable service for millions of people around the world." If that happens, Twitter will turn out to be one of the most important and valuable companies in the world.

While initially leery, legacy news media and their reporters soon learned to adopt and adapt to Twitter. Its speed and brevity are now regarded as among the best ways to break news, not only to the digitally savvy but also through them to the world at large. Dorsey, Williams and co-founder Biz Stone never envisioned Twitter as a source of breaking news, but it quickly became one, as short bursts of text coupled with links to images of disasters provided by ordinary citizens began to spread virally. One pioneer, Portland's *Oregonian* newspaper, presciently began using the service as far back as 2007, when Twitter had just 500,000 users, posting its own links and aggregating other tweets about flooding and road closures during heavy storms then in central Oregon. With legacy media steadily downsizing and shutting bureaus, Twitter came to be viewed as an alternative source of timely, useful news when and where other media were not yet—or no longer—on the scene.

Speed vs. Accuracy
Although being first to report the news is obviously important, reporting it accurately has long been viewed as even more so. The rise of social media, however, has raised new questions about such "old media" values as the balance and interplay between speed and accuracy. Some analysts, such

as Twitter's in-house media strategist Robin Sloan, believe there are both good and bad things about how rapidly news is transmitted on Twitter. "There's no doubt greater speed has a cost," Sloan admits. "But the truth of matter is that we can't slow it down. The news metabolism is speeding up, and social media is now the collective heartbeat we all have. It must be part of the conversation. So there's no real question whether to engage in it or not."

Six months after news of the "miracle on the Hudson" broke on Twitter, another event—the sudden death of pop icon Michael Jackson—brought these issues into even sharper focus and revealed Twitter's speed to be a double-edged sword. In a post on his Technologizer site, headlined "Twitter: The Fastest Way to Get Informed. Or Misinformed," former *PC World* editor in chief Harry McCracken blogged about how he had followed news of Jackson's demise both on television and on Twitter. "When I happened to turn on the TV, MSNBC was still speaking of Jackson having gone into cardiac arrest," McCracken noted. "The (correct) consensus on Twitter was that he had passed away. Impressive proof of Twitter's speed and old media's lethargy, no?"

Yes . . . and no. Like many, McCracken spent most of the rest of the day, once "television caught up with the tweets," watching network coverage of Jackson's death. But when he checked back in with Twitter, the mourning included another dead celebrity, actor Jeff Goldblum. "The sad news had broken that he had fallen to his death while filming a movie in New Zealand," McCracken said.

There was one small problem, however—Goldblum was still alive. The reports of his death, like those regarding Mark Twain more than a century earlier, had been greatly exaggerated. The story of Goldblum's passing was "a hoax created with a tool for creating fake stories about famous people," McCracken wrote. "It took me about 90 seconds of Googling to learn that." Nevertheless the false reports of Goldblum's death spread rapidly throughout cyberspace. As Twain once noted, "A lie can travel halfway around the world while the truth is just putting on its shoes."

Fact-Checking . . . After the Fact
What lessons can be learned from the Jeff Goldblum hoax and others like it? "Part of the reason why information travels quickly on Twitter is that it's not fact-checked. Or more precisely, it's fact-checked after the fact, when people realize the original tweets were wrong," McCracken explained. McCracken also offered some useful cautions and tips for checking the credibility of reports found on Twitter, "If a single person you know and trust tweets something that sounds unlikely, it's more likely to be true than if 500 random strangers tweet it," he said. "But check it anyway." In addition, "If a huge story breaks on Twitter, give the 'old' Web ten minutes to catch up. If neither CNN.com, NYTimes.com, or MSNBC.com has any mention, Twitter probably got it wrong."

McCracken sees a social solution to Twitter's trust issue. "Twitter, or Twitter-like services, will eventually go a

continued on page 129

Interview with Twitter Co-Founder Biz Stone

Stone offers his thoughts on the rapid evolution of the service, as well as the topic of emerging media, trust, and journalism.

ROC: What is Twitter? How would you describe it? Is it a social network?

BS: Twitter is a 24-hour feed of everyone in the world; a soundtrack to our universal film; the Zeitgeist to news on wires. Twitter is social media, but NOT a social network—it's a place where you can zoom in and out on trends and emergent topics. When you think of the entire ecosystem as an organism, that's when it begins to get really interesting. . . .

Twitter is about the idea of an organic approach to communication. We come at it indirectly, organically. Twitter messages only go to an opt-in community, which makes it easier to engage in open conversation. Of course, when a news event happens, we want more engagement. At other times, you can turn it off, as the settings allow user control.

ROC: What are Twitter's uses for journalists?

BS: The news applications surprised us. We noticed in prototypes early on, though, that things like earthquakes led to Twitter updates. The first Twitter report of the ground shaking during tremors in California, for example, came nine minutes before the first Associated Press alert. So we knew early on that a shared event such as an earthquake would lead people to look at Twitter for news almost without thinking.

ROC: Are there advantages to Twitter beyond speed, beyond simply being first with breaking news?

BS: Well, during the earthquake I'm referring to, there was a lot of depth of reporting as well—3,600 separate updates on Twitter, which is the equivalent of a fifty thousand word book in terms of content size. And I'm confident that had the quake been worse, the next step would be in journalists using it to find human-interest stories. (Incidentally, we might also have seen social collaboration activated via the service to help people!)

It's also interesting that Verizon's voice network broke down during the quake, but Twitter's service didn't, because our packet switching technology is more reliable than telephones. But in the end, it's not about technology—it's about the idea of connecting in groups more quickly and efficiently.

ROC: What are some ways journalists are using Twitter?

BS: We were also surprised at how quickly and expertly news organizations—places like the *New York Times*, CNN and so on—began to use Twitter. They just jumped in and impressed us with how they engaged, and their hybrid approach. Reuters, for example, began watching Twitter for trends, and found it worked. We gave help, support, and even our API (application programming interface) to the Reuters Lab people. Then CNN began using us to access information, and to find and create stories. Rick Sanchez at CNN, for example, used both Facebook and Twitter to get real time feedback. . . . And the *Los Angeles Times* took the Twitter feed about the wildfires and put it on their home page.

Another good example is the story of the Twitter user who blogged just one word—"Arrested"—and had the story of his detention splashed instantly to the world's attention, thus leading to his quick release.

ROC: Is Twitter also useful in search?

BS: We are involved on a macro level in documenting events. If you go to search.twitter.com you can discover and cover trends in detail every minute. You could call it "search," but it's really not. "Search" on Twitter is more about filtering results before they hit the Internet—so it's more a kind of filter than actual search.

ROC: Can social media such as Twitter help solve journalism's trust and credibility problem?

BS: We think that social media is largely comparable to the traditional approach, in that credibility is key. In the future, social media tools will help the news media know such things as the location of the person reporting. We will be able to provide a social graph of our users. . . . Can we then triangulate about their credibility via algorithm? We can certainly begin to get very sophisticated on credibility with new tools, and combine that with journalists leveraging open systems such as ours to find and vet crowd sources, story leads, etc.

Looking ahead, I see more sophisticated tools to deal with this issue. A credibility algorithm may be possible one day. Maybe it is even now, as rudimentary as it would still be. Our election feed, for example, was a smart feed. As we go forward and learn more about open systems, we can filter better and thus get more credibility. But filtering is how we get there . . . so one should not rely on social media alone.

long way towards solving this by figuring out how to weight the contributions of the most reliable members the heaviest," he forecasts. "So random people believing everything they hear don't spread falsehoods quite as fast." McCracken concluded with a warning: "Imperfect though Twitter may be, I love it. But I consider it a source of news leads, not news." In other words, trust . . . but verify!

Michael Jackson's death opened another window on how a media system in transition between old and new now handles news reporting and distribution. Writing on the *Tech Crunch* blog the day after the demise of the King of Pop, in a post headlined, "Mainstream Media Still Has Eyes Wide Shut," Robin Wauters decried the fact that few of the mainstream media "dared admit that blogs and Twitter had simply been quicker with spreading the facts than they were." Instead, they claimed it was "old media stalwarts that did the heavy lifting."

The assertion caused Wauters to laugh aloud. "Chest-beating over old media doing the 'heavy lifting' for blogs and Twitter, and being faster in reporting information than those new media when it was exactly the other way around is beyond ridiculous," he wrote. "To me, this whole thing just proves that mainstream media are justifiably freaking out with their eyes wide shut to what's happening instead of learning and adapting to the new age of journalism."

In this new media world, "News no longer breaks, it tweets," as digital analyst and researcher Brian Solis noted in a post on the paidcontent.org site, entitled "The Information

Divide Between Traditional And New Media." In the current era of the real-time web, Solis said, information travels faster than the mainstream media can report it. "Human networks" like Twitter function as virtual news networks, and in the process defeat traditional media in the race to be first.

"We no longer find information; it finds us," Solis observed, since social media "dramatically reduces the time between an event and collective awareness." Trusted messages on Twitter are rapidly re-tweeted by others; their news becomes increasingly prominent and pervasive. The gap between a jet falling in a river and the journalistic reporting of it almost immediately "fills with tweets, updates, and posts as the crowd-powered socialization of information steps in to fill the void."

As Janis Krums demonstrated at the "Miracle on the Hudson," information now moves "with or without the legacy media . . . with far greater speed, reach, impact and resonance," Solis noted. An information chasm has opened between the social media and their mainstream counterparts. Slowed by the time they take "to discern, document, fact check, and publish material information," legacy media loses the race to be first as reporting on social media speeds ahead, "whether or not it is completely or only partially based on facts."

In an effort to narrow the divide, the legacy media increasingly neglects its obligation to discern, document, and fact check. One case in point: In March 2010, as part of a class experiment, a criminal law professor at Georgetown

University Law Center named Peter Tague informed his students that U.S. Supreme Court Chief Justice John Roberts was in poor health and planned to retire soon. Tague did not reveal his source and asked his students to keep the news confidential.

Midway through class, Tague explained he had made up the story. He hoped to illustrate an important point to the lawyers-in-training: even if you receive information from a credible source such as a law professor, it can still be inaccurate. The lesson seemed clear—trusted news and information should be based on multiple sources.

It's an important lesson for all, of course, but especially for journalists. Proof came just half an hour later, when the RadarOnline site reported as fact the rumors of Roberts' pending resignation. The gossip site had picked up a Twitter alert sent by one of Tague's students and promptly posted an "exclusive" speculating on Roberts' health. Later, instead of retracting that report as erroneous, Radar falsely reported that Roberts had changed his mind. In the meantime, Fox News and other legacy outlets broadcast the original, fictitious story.

It no longer even takes a prank, a tweet, and a gossip site to spread false information, however. In January 2011, for example, when Congresswoman Gabrielle Giffords was shot in the head in Tucson, Arizona, several trusted legacy outlets, including CNN, NPR, and the *New York Times*, raced to report news of her death. Although she was in critical condition, Giffords was still very much alive. In today's transitional new media world, sometimes even multiple and supposedly credible sources are no longer believable. . . .

This push and pull between speed and accuracy merely reiterates an age-old tension in news, pitting the urge to get it first against the need to get it right. "The prolonged cycle of journalism and reporting, while slower than the human algorithm that powers the now Web, is still unrivaled, however, by its dedication to discovering, verifying, and reporting truth and fact," says Brian Solis. But in "the race towards veracity, the checks and balances of new media systematically reduce error and filter hearsay and speculation . . . long-standing sources are now slowly losing favor as a destination for revelation."

Since both speed and accuracy are crucial in news reporting, separating truth from rumor and fact from fiction remains essential for maintaining trust. New media such as Twitter offer their own differing forms of checks and balances, which although imperfect, still help reduce error and filter hearsay and speculation. Inaccurate reports such as the deaths of Goldblum and Giffords or the resignation of Roberts crop up periodically; they are soon corrected by the "wisdom of the crowd."

"The scale is so enormous—you can't possibly read 100 million tweets per day!" says Twitter's Robin Sloan. "The puzzle, and our biggest challenge, is how to organize it all, how to separate signal from noise to find the good stuff. So absolutely we need filters—plural."

Sloan believes that although machine learning and

recommender systems powered by algorithms will play an important role in filtering news in the future, human-powered verification will remain primary. "There will not be just one kind of filter but independent, smart, individual voices are very effective. And many of them already on Twitter *are* journalists already," he points out. "Ultimately, we put more trust in humans.

"Brains and human voices are the most important filters," says Sloan. "A computer can't figure it all out! No algorithm is perfect, and no instant filter shortcuts like that will become available. Anyway, the human voice is inherently more trustworthy."

Sloan also feels that legacy media brands still have a role to play, broadly speaking, in the trust equation. "The imprimatur of the *New York Times*, for example, is still quite valuable," he points out. "But even they still have to win authority. The brand power that used to be, say, Walter Cronkite—those days are gone for good—and that's good."

Curation will also play an increasing role in the filtering process in the future, says Sloan. "Big brands will need to go higher up the news food chain and ask themselves what they do that an individual voice acting as a filter cannot do."

Can they, for example, have a broad scope and global footprint, such as *The Economist* magazine has succeeded in creating? Can they establish a presence on new devices and then sell commodities? Can they learn to think not just about the news per se but more about actual products they can build, such as apps and so on?

"Anyone can filter and create news," says Sloan. "Not everyone can create products. Eventually, individuals operating in a vertical content space will become micro brands— and they will beat any more generalized brand competing one to one in that space. In the end, people prize human voices. Twitter is a conversational medium that needs a human voice—even for brands."

Sloan also sees a coming need for the "democratization of correction practices" as well as of reporting and distributing news. "We used to just hear about the news, but now we are all producing and sharing it, now we all have the experience of journalism," he says. "If Twitter and other social media are opening up the process of journalism, we need to show more of process, and not pretend to know it all. Instead let's present our news, share it, ask questions and get feedback. And if we make a mistake, we need to correct it."

Analyst Brian Solis makes a similar point. "We are all in this together, all practicing journalism now in a real-time competition for mindshare, connectedness and earned relevance," he says. "Information is no longer an isolated or individual experience; instead we are connected based on common interests, networked online collaboration and social media." As a recent survey by the Pew Research Center's Internet & American Life Project and the Project for Excellence in Journalism shows, we have become a nation whose relationship to news is becoming "portable, personalized and participatory."

This ability to plug in to social networks "and the

invaluable relationships that define them is where the transformation begins and the journey unfolds," Solis concludes, echoing Robin Sloan. "In the end, we earn the attention, relationships, and audiences we deserve." A new, collaborative journalistic hybrid is beginning to "open new doors to relevance," Solis says. "Connecting to stories and people that propel information beyond the reach of any one network at the speed of the now web."

And Twitter is more "now" than any other media on the web—at least for now! Like the media world that spawned it, Twitter is still rapidly changing. Its creators didn't originally plan for it to become a platform. Evan Williams says, "We launched Twitter sort of as a Model T—it was very basic, but was popular, and it got people excited." Yet it has become one of the simplest, easiest, and most efficient forms of mass publishing ever invented, and Williams says he is still exploring its full ramifications along with the rest of us. Twitter has "lowered the barriers to publishing almost as far as they can go," as Williams told a crowd at a Web 2.0 Summit in November 2010. As a result there are "more voices and more ways to find the truth, then the truth will be available to more people," he believes. "I think this is what the Internet empowers [but] society has not fully realized what this means."

While at the center of this explosion of voices, Twitter is of course just one of many tools of media empowerment. As Web pioneer Dave Winer has noted, the Internet itself is the most powerful tool—not the specific services that run on top of it, such as Twitter and Facebook, which he compares to brands like NBC. These new media brands have become powerful because so many people use them in real-time networked communication.

Along with YouTube, Facebook and a plethora of other new social media, Twitter provides us with a fire hose of news and information—some quite meaningful but a lot that is not. It's up to us to figure out how to sort through it all. Trend-filtering and curation tools such as Storify, built by former AP correspondent Burt Herman), Curated.by (a website) and other new services now enable us to pull various threads and fragments of information and conversation together to begin to make sense of them.

As industry analyst Ken Doctor has observed, news is everywhere now; it comes to us in different forms and different ways, and clearly Twitter is among them. At its best, during events such as the recent uprisings in Tunisia and Egypt, Twitter allows for the true crowdsourcing of journalism—a powerful tool for the pursuit of truth. Some legacy media firms have figured out how to use Twitter and other tools to take advantage of this transformation of the news industry, but most still have not.

The Twitter News Network

Although Twitter remains fundamentally about communication, it is becoming less conversational and more like a networked news organization. The most notable changes include its interactive aspects and embeddable media elements, such as videos, photos, and photo streams. Previously the only components of a tweet had been the text and the link; now we're seeing a shift from text-and-link toward text-and-image, away from conversation and toward news and information.

"Twitter, like blogging did before it, puts the tools of publishing in anyone's hands. And yes, that means the information flowing through the network is not always accurate—hoaxes are a routine part of the stream—but it also means that there are thousands more eyeballs and brains studying those reports than there would be at any mainstream media outlet," as Mathew Ingram noted on the GigaOm site. "The 'people formerly known as the audience' have the tools to become part of the media now, and that is changing our society in ways that we are only beginning to appreciate."

Today's Twitter, "as compared to the Twitter.com of yesterday, is much more about information that's meaningful and contextual and impactful," Megan Garber wrote on the Nieman Journalism Lab site. "Which is to say, it's much more about journalism."

Arizona State University journalism professor Dan Gillmor agrees, saying journalists should view Twitter as a "collective intelligence system" that provides early warnings about trends, people and news. Journalists, he says, should "follow people who point them to things they should know about" and then direct questions back to them to do better reporting. Gillmor recommends setting up keyword searches and understanding "hashtags," Twitter-speak for a group of tweets about the same subject or event, indicated by a # sign and topic word (such as "#occupywallstreet.")

"We go to legacy newsrooms often," Robin Sloan told me. "It's noteworthy that when I look around, Twitter is open on lots of desktops. I find it interesting and meaningful that journalists of all types find that using Twitter is close to the type of work they've always known—helping them keep track of broad, distributed network of sources, for example.

"On the other hand, Twitter has also transformed the way journalists think about their work," Sloan contends. "Now they get news ambiently, if you will. And there is a large place for professional journalists in Twitter. In a sense, it makes them more valuable than ever as they curate and filter and then present information back to us."

Brian Solis once referred to Twitter as TNN, the Twitter News Network, since "it consistently beat traditional media in the race to report relevant news and trends." Company co-founder Biz Stone, it turns out, sees the future much the same way. In November 2010 Stone told the Reuters wire service he is eager to find a way to harness the vast quantities of information shared on Twitter to create a news network. Such a network, Stone said, would not necessarily be run by Twitter itself but could be in partnership with several legacy news organizations such as Reuters. "From the very beginning this has seemed almost as if it's a news wire coming from everywhere around the world," he told Reuters Television at a technology event. "I think a Twitter News Service would be something that would be very open and shared with many different news organizations around the world."

"The train to the future is moving down tracks, and there are plenty of people onboard with no interest in news and information. So it's especially important for those who do care about it to get on that train and push their values," concludes Twitter's media strategist Robin Sloan. "Everything is changing so rapidly now that the future will make the current transition look like a picnic. It won't end with Twitter. So legacy media can't just 'learn Twitter' and be done with it. The key is curiosity—to successfully navigate through it, you must remain curious."

8.

Google Loses Its Buzz

It's of course impossible to consider the disruptive effects of the digital information revolution without examining the role of Google. Founded near the turn of the century, the company rocketed from start-up to the world's most powerful and profitable Internet firm in just ten years. Google's vaunted search engine completely revolutionized the way the world gathered information. Its impact is now so great that some observers—such as self-described "Google fan boy" Jeff Jarvis in his bestselling *What Would Google Do?*— hail the firm in terms usually reserved for deities. Other company chroniclers, such as Ken Auletta in *Googled: The End of the World as We Know It*, simply and accurately say that corporations everywhere have been "googled."

Yet for all its success, by the end of the decade the company faced a host of problems. Google's awesome power and reach proved to be a double-edged sword; competitors and regulators alike assailed it for a series of antitrust and privacy violations and began demanding remedies. At the same time its Web supremacy came under attack by new competitors such as Facebook and Twitter, as Google lagged behind in what was fast becoming the most engaging and potentially lucrative online phenomenon of all—social media.

Google's failure in the social networking arena wasn't the result of a lack of desire. As far back as 2003 its executives had tried unsuccessfully to purchase the leading social network Friendster; the next year the company launched its own social network called Orkut. Although the service became popular in India and Brazil, where until recently it remained far ahead of Facebook and other competitors, Orkut never caught on in the United States. As a result, savvy investors such as venture capitalist Fred Wilson complained that Google had "missed the whole social networking thing. Facebook beat them to that."

The Antisocial Social Network

Alarmed at the rise of social media and the attendant perception that they had missed something important in the evolution of the Web, Google executives tried again in 2010 by adding what they called a "social networking feature" to the popular Gmail service. Called "Buzz," this new tool for sharing personal information allowed users to post status updates, share content, and read and comment on posts in much the same way they could on Facebook or Twitter. The decision to piggyback the new network onto Gmail, which already had more than 150 million active users, was meant to vault Buzz immediately into the top ranks of social networking sites.

This long-anticipated "Google approach to sharing," as company flacks phrased it, was clearly an attempt to restore Google's reputation as the most innovative and important force in the digital information space. But would Buzz actually "change the way businesses communicate around the world," as Google's director of product management Bradley Horowitz crowed? Was company co-founder Sergey Brin correct in his claim that Google would reinvent social networking in much the same way the company had reinvented search a decade earlier?

The answer soon became obvious, as privacy and trust concerns caused Buzz to flop almost immediately. The hasty decision to use Gmail as the launch pad for building a social network meant that everyone in its user base was instantly enrolled—like it or not. Evgeny Morozov, whose writing focuses on the impact of the Internet on repressive regimes, became one prominent critic upon having his private contact list—one that government officials in authoritarian states might find of great interest—made public without his knowledge or consent. Many others were equally quick to complain about what they too saw as Google's violation of trust and invasion of privacy. Whereas they had freely chosen to join platforms like Facebook or Twitter, none had been asked or granted permission for Google to raid their personal Gmail lists in order to set up Buzz. As previously private names and addresses suddenly become very public, a firestorm of negative reaction followed. "Angry tales were being told of people's contact details and other information being passed on to psychotic and abusive ex-husbands," the *Times* of London reported. Using Buzz, a complaint filed with the Federal Trade Commission warned, could easily risk the exposure of "estranged spouses, current lovers, attorneys and doctors."

Google executives moved swiftly to contain their self-created crisis, rolling out a "privacy reset" that dropped the automatic sign-up and offered clearer instructions on how to opt out and keep messages private. "Shortly after launching Google Buzz, we quickly realized we didn't get everything right and moved as fast as possible to improve the Buzz experience," product manager Todd Jackson explained in an apologetic blog post. "Offering everyone who uses our products transparency and control is very important to us."

Although the abrupt fixes did make it clearer how information was shared and simplified the process for blocking or following other people, the changes did not go far enough. Nearly a dozen members of Congress expressed concern over claims that Google Buzz "breaches online consumer privacy and trust," and asked the FTC to investigate. The Electronic Privacy Information Centre alleged that Buzz violated consumer protection law and was "deceptive," as the service became the subject of a class action lawsuit.

The controversy left the search giant "fighting a rearguard action in the latest skirmish over privacy on the internet," the *Times* of London added. In fact, Google had stumbled so badly that Buzz rapidly became a laughingstock, widely derided as an antisocial social network. "We made some mistakes and we accept that," the firm's head of communications, Peter Barron, admitted. "But if you look at the way we responded, I hope people will see that we reacted quickly to those criticisms and made significant improvements. We take privacy very seriously and build privacy features into all our products based on the principles of transparency, choice and user control," Barron added. "At Google, users' trust is all we have."

"They have to do something"

Although intended to give Google a stronger foothold in social networking, Buzz had just the opposite effect. Arrogance and an inability or unwillingness to listen had led executives to violate trust and harm privacy relationships with Gmail users—with dire if predictable results. The leading technology blog Mashable called Buzz the "biggest tech flop of the year," while noting that, "With Google's biggest attempt at social now a mere afterthought, nothing stands in Facebook's way." Perhaps worse, the bad buzz about Buzz only "served to reinforce a bigger narrative about Google," as the *Financial Times* reported. The company was perceived as using its power as "a blunt instrument to muscle its way into new markets—and it is not too concerned about whose feet it treads on in the process."

Individuals like Evgeny Morozov were left feeling like pawns in a corporate chess game played among giants. Because Google "is losing traffic to Facebook and Twitter," he complained, "They have to do something." Although

Google executives apologized and changed policies, severe damage had already been done to the reputation of a company with the boastful corporate motto "Don't be evil." Competitors seized the opportunity to attack; Apple founder Steve Jobs, for example, chortled that Google's motto was just "a load of crap!"

While conceding they hadn't communicated with Gmail users well or sufficiently in advance, Google's leaders maintained that their mistakes were made in good faith. "You can't incubate these kinds of products in a Petri dish," Bradley Horowitz complained. But critics such as Marc Rotenberg of the Electronic Privacy Information Center shared Morozov's suspicion that the firm's actions were actually driven by a deeper corporate agenda. "The way they could compete was to enlist all the Gmail subscribers," Rotenberg noted. "That's a very clear corporate decision."

Worse Than a Ghost Town
At the end of June 2011, Google surprised the tech world again by making yet another attempt to compete with social media leaders like Facebook and Twitter: the Google+ project, heralded as "Real-life-sharing, rethought for the web." Senior Vice President of Engineering Vic Gundotra ballyhooed the new social network's launch on the company blog. "Among the most basic of human needs is the need to connect with others. . . . Today, the connections between people increasingly happen online," he reminded us. "Yet the subtlety and substance of real-world interactions are lost

in the rigidness of our online tools. In this basic, human way, online sharing is awkward. Even broken. And we aim to fix it."

Given Google's abysmal track record in social, Gundotra's claim seemed bold to many. Yet if any company were able to go head-to-head with Facebook, it would be Google, with its massive installed user base. "You and over a billion others trust Google, and we don't take this lightly," Gundotra noted. "In fact we've focused on the user for over a decade: liberating data, working for an open Internet, and respecting people's freedom to be who they want to be.

"We realize, however, that Google+ is a different kind of project, requiring a different kind of focus—on you," Gundotra concluded. "That's why we're giving you more ways to stay private or go public; more meaningful choices around your friends and your data; and more ways to let us know how we're doing. All across Google."

Upon further examination, Google+ revealed itself mainly to be an amalgam of several existing services already used by many. The idea as explained on was simply to do them better. At first it appeared as if Google's new social strategy might work; the initial adoption rate set records. It took just 24 days for Google+ to reach 20 million users, for example, whereas it had taken Twitter 1,034 and Facebook 1,153. Google+ came zooming out of the gate as one of the fastest growing websites in history, and the early adoptions had some analysts convinced Facebook was finally facing a serious competitor.

Within a few months, however, Google's latest attempt at socializing itself went from boom to bust; traffic to the Google+ site had plummeted by sixty percent. Media analysts now wrote eulogies, with one declaring the service "worse than a ghost town," and another noting, "[T]here are a few things that are slightly better [than Facebook and other existing social media platforms,] but what's really making a huge difference? You know, that's the problem. There's nothing really groundbreaking."

Part of the problem was the fabled network effect: with Facebook and Twitter already commanding huge attention, many questioned the need for another entry in the social space. "Is it possible that G+, at the moment, is simply a social media step too far?" asked Dan Reimold on the PBS Mediashift site. "Are there only so many daily destination-and-connection sites a person can invest time and effort overseeing?"

Paul Tassi answered for many when he wrote on the Forbes.com website, "The fact is, very few people have room to manage many multiple social networks . . . since there is only so much time in the day to waste on the Internet. Add in Google+, effectively a duplicate of Facebook, and there just isn't space for it." Calling Google+ "a failure no matter what the numbers may say," Tassi described clicking on his newsfeed there, only to see "tumbleweed blowing through the barren, blank page. It's a vast and empty wasteland, full of people who signed up but never actually stuck around to figure out how things worked in this new part of town.

One simple click takes me back to Facebook, and my wall is flooded with updates and pictures from 400+ friends. This just isn't a contest, and it never will be."

"I am writing to second Tassi's declaration: Google+ is dead," pronounced Reimold. "At worst, in the coming months, it will literally fade away to nothing or exist as Internet plankton. At best, it will be to social networking what Microsoft's Bing is to online search: perfectly adequate; fun to stumble onto once in awhile; and completely irrelevant to the mainstream web."

"Why manage two different social networks where the only difference is cosmetics and a few bug fixes? Why post a status update or album to Plus where ten people will see it, when you can do so on Facebook where it will catch the eyes of 500 friends instead?" Tassi asked. "Google just should have known better. No one is going to scrap a social network they've spent 8 years building up to start over from scratch for one that offers only a few minor improvements. To compete there needs to be something put forward that's truly revolutionary, and tech companies half-heartedly copying each other is not going to cut it and can't masquerade as true innovation."

Newspaper columnist Rainbow Rowell summarized the problem well when she wrote, "It's a not-vicious-enough-to-be-interesting circle: Nobody posts on Google+ because nobody posts on Google+. My Google+ home page is worse than a ghost town. It doesn't even feel haunted."

Google Has Become a Jungle

Social media threatened Google in many ways, including by mounting a full-frontal assault on search itself—the company's existential core. Critics began to carp, "that the digital age's most mundane act, the Google search, often represents layer upon layer of intrigue," and wondered aloud if the whole notion of search rankings had become "an outdated system." Google's founders, as Ken Auletta noted in *Googled*, were "acutely aware that search is still fairly primitive."

"The Silicon Valley company built its business on the strength of algorithms that yield speedy results," the *Wall Street Journal* noted in March 2011. "Last month, Google acknowledged it 'can and should do better' to beat back sites that 'copy content from other websites' or provide information that is 'just not very useful' but are ranked highly anyway."

"I've never seen Google be attacked on the relevancy of their results the way they have these past couple of months," Danny Sullivan, editor of the influential *Search Engine Land* blog, told the *Journal*. Vivek Wadhwa, a visiting scholar at the University of California, Berkeley, added that his students had trouble finding basic information on Google. The problem was a familiar one. "Content on the Internet is growing exponentially and the vast majority of this content is spam," said Wadhwa. "Google has become a jungle."

In response, the company tried to tweak its secret search formula in an effort to improve the quality of its results. Google senior engineering manager Amit Singhal said numerous new "signals" had been incorporated into the search algorithm. Many were human, or social, in nature, such as indications of "how users interact with" a site, he said, or feedback from hundreds of people hired to evaluate the changes. The "human raters," Singhal said, were asked to look at search results and decide whether they would give their credit card number to a site or follow its medical advice.

Despite the tweaks to its algorithms and addition of human raters, however, one incontrovertible trend remains and it is the most threatening of all to Google's hegemony: "social" is beginning to replace "search" as a primary means of finding credible news and information. Rather than use Google's "pull" model, which offers numerous links in response to a single query, social media allow for a more efficient "push" model, where friends and followers deliver fewer but more targeted and trusted answers.

Other new search approaches coming from social models—such as "real-time" and "vertical" search—also worry Google executives, while newer, location-based networks such as Foursquare and Gowalla, threaten the Google advertising model by offering an even more precisely targeted approach to consumers. Even Ken Auletta, who like Jeff Jarvis is generally a Google enthusiast, concedes, "Social networks might pose a threat to Google search." Auletta says he fear the future effect of social media on "one of the world's most trusted corporate brands."

Leading industry analyst Om Malik of GigaOM is among those convinced Google should fear social media, since we are now in the midst of "a major shift in the web and its core ideas—a transition from data web to the context web." Our newly networked lives have become overloaded; the recent exponential increase in the *amount* of information available is inhibiting our ability to find *credible* and *relevant* information. It's time we move, says Malik, from the "sell-search-and-consume methodology that has become part of our basic Internet behavior and turned Google into a gazillion dollar company" to a new world of social search.

Unlike Google's methodology, which relies on web links to rank results, social search will shift power to individuals using social tools to express their opinions. In other words, it will democratize and humanize the search process by using friends and followers instead of algorithms to provide context to and filters for our ever-expanding amount of information.

"The company that is most impacted by these developments is Google, the shining example of the Data Web," as Malik notes. "By deploying its awesome infrastructure and massive computer resources, Google has enjoyed an advantage over all its search competitors." That advantage is now disappearing, as the manner in which we find and use information on the Web is being transformed. Despite its endless efforts to play catch-up with Facebook, Twitter, and other social media—witness the September 2011 public launch of Google+ and the October 2011 official demise of Buzz—

Google simply may be incapable of doing so. The company "lacks the DNA that would mark it as a social entity," says Malik, and it has never "been comfortable dealing with the 'social' or 'people' web. Look at any of their offerings—they have the warmth of a Soviet bunker."

Antisocial *and* Antitrust?

Even as Google executives struggled to cope with the social media challenge, other threats awaited them. With its massive brand appeal, consumer watchdogs considered the firm "an ideal target" and urged antitrust agencies on both sides of the Atlantic to scrutinize its activities much more closely. As the company expanded into businesses beyond search and advertising, regulators began to take a closer look and launched investigations aimed at ensuring that the search giant could not act unfairly toward either consumers or competitors. Once again issues of privacy and power came to the fore, as concerns over an antisocial social network banged into questions about trust and antitrust.

In Europe, where Google is particularly dominant in search and online advertising markets, competitors complained to politicians and regulators about the ways the company divided advertising revenues and controlled what appeared in its search results and in Google News. Privacy groups raised concerns about the company's possession of huge amounts of users' personal data, and Germany's Justice Minister, Sabine Leutheusser-Schnarrenberger, expressed fears that the firm had accrued too much power and information about citizens and warned that it could face legal action if it did not become more transparent.

Industry groups accused the company of manipulating search rankings to favor its own content and suggested that the European Commission open a broad antitrust inquiry. Newspaper representatives seeking a larger share in the information market called for enforcement of "fair competition among all players." Google's Street View mapping service with its intrusion on privacy, its free integration of Google Maps on other websites, and its project to digitize books all disturbed European Union officials, who opened a formal investigation into allegations that "Google Inc. has abused a dominant position in online search, in violation of European Union rules." They also questioned such practices as "whitelisting," or adjusting the results of Google's algorithms to favor certain sites in search results. The issue had specific relevance in the United States as well, as antitrust investigators in Texas began seeking evidence about the "manual overriding or altering" of search result rankings.

Google's sheer size, market dominance, and apparent ambitions made it a natural target for regulators. The company responded by rapidly increasing how much it spends on political lobbying—it is now among the biggest Internet and technology lobbyists in Washington—and keeping executives busy explaining the "six principles of competition and openness" to a range of policymakers in both the US and Europe.

Partly in response to the setbacks, Google underwent a

management shakeup in early 2011. Attempting to rediscover the company's roots as a start-up, co-founder Larry Page took over as chief executive. Eric Schmidt, the tech industry veteran brought in a decade earlier to provide so-called adult supervision to Page and Sergey Brin, the company's other co-founder, stepped aside and became Executive Chairman. Brin would concentrate on new products while Schmidt would focus on "deals, partnerships, customers and broader business relationships," as well as government outreach, including the fight against regulators.

The changes in the executive suites came at a time when the company had already lost much of its entrepreneurial culture and become a slower-moving bureaucracy, in sharp contrast to Facebook, Twitter, and other fresh young competitors. Echoing Mark Zuckerberg, CEO Larry Page told the *New York Times*, "One of the primary goals I have is to get Google to be a big company that has the nimbleness and soul and passion and speed of a start-up."

In announcing the changes, Eric Schmidt said, "When I joined Google in 2001 I never imagined—even in my wildest dreams—that we would get as far, as fast as we have today. Search has quite literally changed people's lives—increasing the collective sum of the world's knowledge and revolutionizing advertising in the process. And our emerging businesses—display, Android, YouTube and Chrome—are on fire." He heralded Google's financial success and maintained that the latest quarterly results proved that the "outlook is bright."

Schmidt was correct in noting that Google remains immensely powerful and hugely successful, of course, with more than 24,000 employees and a market value of $200 billion, up from $27 billion when it went public in 2004. But Schmidt didn't mention social media at all, or the fact that Google's vast share of the online advertising market is now also threatened by antitrust inquiries. The unspoken fear within the company is that it well could begin to resemble Microsoft, the once-dominant technology company that now is widely perceived as stodgy and seems past its prime. Indeed, for all its success, Google is no longer considered by many engineers as the most desirable place to work in Silicon Valley; the new generation of start-ups has stolen its thunder. In recent years, Google has lost scores of top programmers along with a string of high-profile senior executives, most notably Sheryl Sandberg, now Chief Operating Officer at Facebook.

Even as Google+ continued to crash and burn, renewed concerns about Google's monopolistic position in its core business of search advertising returned to the fore. In September 2011 Google's antitrust issues had their most public airing yet when Schmidt was called to testify before a U.S. Senate panel. "Google is a great American success story," said Senator Richard Blumenthal, Democrat of Connecticut, a member of the Senate antitrust subcommittee, "but its size, position and power in the marketplace have raised concerns about its business practices, and raised the question of what responsibilities come with that power."

As the *New York Times* reported, "though the company and the times are different, there are echoes of a hearing before the same Senate body, the Judiciary antitrust subcommittee, thirteen years ago and the last sweeping antitrust investigation of an American technology powerhouse, Microsoft. Later, the federal government, joined by twenty states, filed suit against Microsoft." Like Microsoft before it, Google is now under intense scrutiny for dominating in one arena—in Google's case, search—and then expanding into numerous others, where it competes with the same companies that it indexes in its search engine. Is it using its powerful position to give its new businesses an unfair advantage?

In testimony to the Senate panel, Schmidt described a company facing fierce competition on many fronts and emphasized the open Internet, where consumers can easily switch to competing services. The Senate hearing had no direct consequences for the other investigations already under way in the United States and Europe, although an ongoing and broad-reaching inquiry by the Federal Trade Commission very well may. By spotlighting Google's potential threat to consumers, competition, and industry innovation, however, the hearing could conceivably turn policy makers and public opinion against the company.

With investigations by European, Asian and U.S. regulators continuing, the last thing that Google executives need is to get bogged down in a lengthy and costly antitrust battle of the kind that cost Microsoft billions in the 1990s. Yet the possibility is very real, as evidenced by the Germany Justice

Minister's remark that, "All in all, what's taking shape there to a large extent is a giant monopoly, similar to Microsoft." After noting that the search advertising market has become "the fundamental economic engine for content online" and "the gateway to content on the internet," Microsoft's general counsel Brad Smith observed, "[W]henever you have a company that has more than a 90 percent market share in a key market, it is inevitable that people will have questions to ask. We say that with some experience."

Whatever the reality, the perception of Google's primacy and purity clearly has changed; even top executives like Eric Schmidt have worried aloud whether, "The halo is off." The botched launch of Buzz, the patchwork fixes to Search, and the non-essential feel of Google+ each shone new light on Google's vulnerability to the emerging social media. They also intensified suspicions that, like Microsoft and other technology giants before it, Google was deliberately trying to use its dominance in one area to increase its presence in other new and important markets and, as the *Financial Times* reminded us, "it is not too concerned about whose feet it treads on in the process."

Any company's failure to communicate adequately to customers has an inevitable credibility cost—a lesson Facebook executives also learned when facing their own outcry from dismayed users and privacy interest groups. The clear message: if Google wants to keep the trust of its users, its executives will have to work much harder to convince us they truly have the best interests of consumers at heart and not just their own. But antagonizing previously loyal users is only one of Google's problems; its actions also put at risk trust relationships with business partners and, perhaps most importantly, regulators. If Google fails to address these concerns adequately—and soon—even the world's most powerful and successful Internet brand could find itself toppled from its perch.

Despite their efforts to maintain dominance in search while still playing catch-up in social, Google's founders face a new and growing credibility gap. "The problem is that while everyone knows Google's corporate motto is 'Don't be evil,'" as the *Times* of London put it, "Not everyone believes it anymore."

9.

The New Breed of Media Researchers

A new wave of research into emerging media, information delivery, and web credibility, spurred by a new breed of academics, is upending the previously accepted conventional wisdom that online social networks tend less to persuasion and more to polarization, fragmentation and reinforcement of prior beliefs. The emerging social media, these researchers suggest, possess certain unique characteristics that enable them to function well as credibility content filters.

Northwestern University's Eszter Hargittai, whose study of young Internet users revealed that most have a frightening lack of knowledge as to how brands like Google actually operate in the information sphere, is among those in the forefront of this movement. Along with BJ Fogg, Paul Resnick, Miriam Metzger, R. Kelly Garrett, Cliff Lampe and others, Hargittai believes social media now play a powerful role not only in distributing information but also in filtering it for trust. Their pioneering work, along with that of colleagues such as Danah Boyd and Judith Donath, both of Harvard's Berkman Center for Internet and Society (see

Chapter 10), also suggests that online social media networks introduce new elements of trust to the delivery of news and information.

Like Hargittai, Miriam Metzger of the University of California, Berkeley believes "social media can definitely play a role in trust filtering." Metzger, whose research centers on Internet credibility issues, says, "There is an interesting phenomenon going on, where under certain circumstances, new media can actually be perceived as *more* credible than traditional media." As one example, Metzger points to reporting about wildfires that had threatened her own neighborhood in California. "During the fires here, the news and information accessible from local social media such as Twitter, Facebook and blogs was more relevant, reliable and current than that on local television, which was broadcasting outdated official press releases," Metzger recalls. "The traditional media was not perceived as useful. Meanwhile, the new interactive media was getting people information they really wanted and needed, in real time."

Ohio State University's Kelly Garrett, whose own social network research has convinced him "the filtering thesis sounds correct," echoes Metzger's remarks. Much of Garrett's work focuses on the issue of "selective exposure," whether or not people prefer to receive information that reinforces their opinions and to avoid information that challenges them. Garrett's findings have important implications, as the abstract of one of his studies notes, "for individuals' exposure to cross-cutting political ideas in a contemporary news environment that affords an unprecedented level of choice." His research shows that "there is no evidence that individuals abandon news stories that contain information with which they disagree."

Our informal Internet communications channels "are certainly expanding and becoming a more important part of the media diet," Garrett says. "Twitter users report they are getting news from more eclectic groups than previously, for example. In the past, credentials were much more important. Now they are being supplanted by 'crowdsourcing,' or as in the case of Wikipedia, by a relatively small group of people with the skills they need and the time to do it."

Most recently Garrett's work has centered on the truth of political rumors. He says his research shows "the glass is both half full and half empty, in that while there are definitely more rumors than ever circulating online, at the same time people are also more likely to access Facebook and other social networks." Ironically, Garrett says the very things that are allowing people to have more access to false information—the Internet, social networks, email from friends and family—also afford more access to tools that help them check its credibility.

"The problem is not, as Eric Schmidt claims, that the Internet is some information cesspool," says Garrett. "Brands are not a solution to friends sending emails with inaccurate information. . . . Brands are not really affecting belief." Garrett says instead that social networks themselves increasingly act as information filters, but "the question is exactly how to craft them to make a positive difference. If social networks offer exposure to diverse groups, the diverse views should act as a foil to rumor dynamics.

"Brands do matter to some extent," Garrett concedes. "The sources people rely on do shape them; there is persuasive power in brands as filters." But he believes, "Getting information from a friend is not so different, however, although it may be more weighted because of the issue of trust."

Different Folks, Different Strokes

Like many of this new breed of researchers, Garrett is convinced, "People use different filters in different ways to assess credibility online. When 'everyone' knows something, we find it suspect. If 'everyone' says something is great and 'we all agree,' that's a problem!" Ultimately, he says, our trust and information crisis is one of faith—and thus of legitimacy.

"With little faith in institutions and less trust in the mainstream media, what some once saw as a solution is now worrisome to most of us. And where once there were just a few brands and gatekeepers that were trusted, now there are so many sources, and we don't all trust the same ones," Garrett says. "But there are plenty of legitimate information sources out there and a wide variety of ways to crosscheck what they tell you. With all the concern about the legitimacy or illegitimacy of news sources, people now need to figure it out for themselves, but there's a crisis of not knowing what to believe or what proposed filter is best." As a result,

"Everyone is looking for a set of metrics or set of shortcuts for belief, and this could prove to be dangerous," he believes.

"People want to make up their own ways of trust assessment," Garrett concludes. "Now at least there is an infrastructure to facilitate this. These outlets didn't even exist before." As a result he is optimistic despite the half-empty glass. "The evidence is mixed and more research is still needed," Garrett cautions. "But I believe we are en route to having a more knowledgeable and engaged society."

Assistant professor Cliff Lampe of the Department of Telecommunication, Information Studies and Media at Michigan State University collaborates on research with a team at MSU, as well as from time to time with Garrett. "Social networks represent a sea change for online interaction," says Lampe. "Once, the Internet was a way to free yourself of earthly bounds. Now social networks such as Facebook facilitate a greater interplay of offline and online relationships." To Lampe, an offline relationship of trust between people increases the likelihood that information delivered by them online will prove to be credible. "It works like this," he explains. "If someone I like—a trusted friend—sends it, I will tend to trust the information."

Lampe is among those who are convinced that online social networks such as Facebook offer unique value for their ability to foster "looser but more extensive social connections, hence giving us more exposure to other viewpoints." Facebook, which has been studied extensively by his team of researchers, is "not useful for close friends and family, but for a larger, more dispersed set of connections, which create more diversity and change the social dynamic," says Lampe. "So it's not surprising that Facebook and other online social nets are being used more and more as news filters, with a beneficial impact on both political and social engagement."

Word of Mouse

Paul Resnick, professor at the University of Michigan's School of Information, is another leading communications researcher. In a 2004 paper that specifically explored social capital and information and communication technologies (ICTs), Resnick noted that weak social ties, "those personal connections that involve less frequent interaction and less personal affection, are especially productive . . . because they provide bridges to broader reservoirs of information." In his paper Resnick listed the media as one of several "areas ripe for transformation," and said, "[W]hen choosing media . . . people are increasingly turning" to word of mouse to replace the word of mouth of friends.

Years before they took place, Resnick foresaw the huge changes coming in the media environment based on the use of emerging media and social networks. "The news industry may be poised for a major transformation if more and more people begin to rely on advice from distant acquaintances . . . to monitor the news and form opinions," he noted, adding coincidentally, "electoral politics may also be poised for a major transformation." In fact, Resnick even seemed to forecast the essence of Barack Obama's successful 2008 presidential campaign (see Chapter 11), predicting, "We could see a return to grassroots political organizing for both presidential and local campaigns. Rather than an old-style ward organization, however, we should expect to see a looser network, with information sharing and mobilization of co-ordinated action mediated by ICTs."

In an interview years after his study of social capital and information technologies, Resnick agreed that the ties facilitated by online social networks are "qualitatively different" than those of their offline, real world counterparts. "Does that necessarily lead to more diverse exposure?" he asked rhetorically. "I would guess the answer is that there is some; yes, the technology and tools they offer could create something different."

The work of Resnick and other researchers now delving into this field fly in the face of many previously accepted notions about social networks, the Internet and how they fit together. Until recently, the consensus position in academia held that the Internet mostly serves as an echo chamber that reinforces already-held beliefs, and as such only further polarizes an already partisan nation. Author and professor Cass Sunstein, a Harvard University law professor and Obama Administration official, articulated this view in works such as *Republic.com* and *Republic.com 2.0*, wherein he concluded that, rather than helping to open minds and expose us to an unbiased array of unexpected viewpoints and useful information, the Internet actually causes us to become more close-minded.

To Sunstein, when we go online, we "personalize" the news we receive and select only the kind of news and opinions we care most about, thus filtering out exposure to different concerns and political opinions and preventing a truly democratic conversation. He believes the inevitable result of sharing information through social networks is polarization and intolerance, and that when liberals or conservatives discuss issues such as affirmative action or climate change with like-minded people, their views quickly become more homogeneous and more extreme than before the discussion. Nicholas Negroponte of M.I.T. calls this supposed trend "The Daily Me" (see Chapter 12).

Harvard professor Robert Putnam, a leading expert who describes social networks as "elaborate collaborative filtering systems," seems to share this "polarization-not-persuasion" fear. In his best selling book *Bowling Alone*, he traced the decline of social networks in late twentieth-century America, and thus of "social capital." To Putnam, social capital "refers to social networks, norms of reciprocity, mutual assistance and trustworthiness." Ultimately, he says, one distinction "among the many different forms of social capital" is especially important: the fact that "some networks link people who are similar in crucial aspects and tend to be inward-looking—bonding social capital. Others encompass different types of people and tend to be outward-looking—bridging social capital."

In *Bowling Alone*, Putnam looked exclusively at face-to-face social networks, such as bowling leagues, and had nothing to say about the emerging online social networks and how they may differ from their predecessors. Later, in the conclusion to another book, *Democracies in Flux: The Evolution of Social Capital in Contemporary Society*, he did allude to the possibility that social capital has not simply declined, but may instead have taken new forms amidst "a generational shift away from some sorts of associative activity . . . towards other sorts," such as Internet communications and "new social movements." Yet Putnam is far from optimistic, arguing that, "there is reason to suspect that some fundamental social and cultural preconditions for effective democracy may have been eroded in recent decades, the result of a gradual but widespread process of civic disengagement."

Not The Second Coming

Asked in a 2011 interview if he agreed that online social networks were uniquely different from the real world offline ones he examined in *Bowling Alone*, Putnam said, "I still don't know the answer, although we are a decade into the Internet and eight years into online social networking. The research is still in its infancy, still at the polemic stage, where people stake out positions in a largely data-free way." He added that he believes that "the research to date has mainly achieved ruling out the extremes. This is not the Second Coming, nor is it killing off offline social networks." But he conceded, "We do know that the net effect of the Internet and online social networks is probably not bad. That is to say, a virtual community is not a 'disaster' as some used to think it would be or was."

Putnam, who once taught Mark Zuckerberg's roommate at Harvard, thinks the original Facebook developed there exclusively for the use of the Harvard community was a "wonderful alloy" useful for keeping in touch with real world friends. "It made a very positive contribution, and yes, it *was* different from truly face-to-face social networks." But he sees the newer, more open Facebook to be "half-and-half, an alleged social network," and thinks that it "used to mirror real life more than it does now." Putnam says Facebook's move away from being a closed community led it to become to a "different, non-alloy type of network," adding that, "I know what 'friend' means in real life, but I don't know what it means any longer on Facebook. Now we have 'air quotes' friends instead of 'real' friends."

Still, he says he believes in the possibility and promise of the Internet and its online social networks. "I am open to the view that the range of political views, for example, that is available online is greater than that in the offline world. And it's hard to imagine we can fix problems outlined in *Bowling Alone* without the Internet," he told me. "In some respects at least, we need the Net to fix things, to make it possible to strengthen and create new ties among people. We need to create more alloys and ways to use the Net, like the telephone was used in the past, to strengthen and deepen previous ties, so the Internet might actually help revive social networks in America."

But Putnam thinks the jury is still out as to online networks' utility in helping us find trustworthy, credible news and information. "I don't know the answer about a news alloy," he said. "That's still a prominent question."

The crucial thing is that it not be set up simply as an information transmission but that it is based instead on accountability—or obligation, if you will. Can a purely Internet-based social network produce joint reciprocity? The underlying important feature is does it transmit obligation, as well as just news and information? Will the evolution of obligation lead us to 'tastemakers' and curation? Twenty years from now, will we be falling into an abyss?"

To Putnam, a "News/Trust Alloy" might incorporate friends and followers, along with learning machines, algorithms, and recommender systems, brands, and influentials, tastemakers and curators. The key, he believes, is accountability. "Filtering of any sort only works as you suggest if there is a system designed to produce reliable information by sanctioning bad information," Putnam told me. "In the news arena, there are tons of reasons for homophily. So ultimately, insofar as using the Internet involves a trust accommodation of some sort, sure, online social networks can function that way, just like in the real world, at least in principle. I'm still not sure the Net provides evidence of bridging capital per se. We don't have the evidence yet, and need more studies to establish that."

Tom Sander is Putnam's colleague and co-researcher, and Executive Director of the Saguaro Seminar: Civic

Engagement in America at Harvard's Kennedy School of Government. Sander says that he and Putnam agree that online social networks, with their more extensive but weaker ties, "have a higher opportunity to create bridging social capital." But he also cautions against romanticizing online communities, warning, "Just calling something a community doesn't make it one. This all needs to be empirically tested."

Nonetheless, Sander believes, "There is no doubt that they can function as a social filters. In this context, news travels faster. There is a low transaction cost and high speed of distribution. In principle, then, a social network for news could work . . . and in principle you could weight 'trust' from friends on your feed over time. There is nothing implicit in social capital/social networking theory that says this *couldn't* happen. It all depends on the creation of bridging capital."

Yet many questions remain, according to Sander. "One problem with filters is that people often apply partisan lenses to them. Do we live in an era of partisan truth?" he wonders. "Or can we uncouple approval from facts? The simultaneous challenge and promise of online communities is that both the entry and the exit are so easy. The cost of being untrustworthy on the Internet can be low because it is so easy to exit 'the community.'

"Social networks can make people more media literate, but ultimately the real challenge is the need for someone to vet the information somehow," says Sander. "Our younger cohorts are certainly using the technology of social networking more and more—that is empirically true—but what are the consequences of this use for social capital? Technology will have to be part of the solution to the credibility issue, but we are skeptical about the actuality of it at this moment.

"Is the Net more of a utility, like a mobile phone or a television, or is it a tool for social change?" Sander asks. "How inclusive are online social networks? They are a valuable means to spread reputational information—but do the people there have expertise? And if so, are they willing to share it?" Finally, he posits, "The real question is even more basic: Will credible information be spread via the Internet?

"That depends on the structure of the social networks there," he says in conclusion. "Determining *that* is the challenge now facing researchers." Although he thinks "the Web is potentially the seed of the solution" to journalism's credibility problem, Sander also warns, "[C]ollectively we need to do a lot more research."

Kelly Garrett agrees the empirical question is whether or not social networks will become more diverse by virtue of technologies like Facebook. "Yes, people will *hear* more diversity—but will they build more diverse social networks?" he asks. "Will it be back to the future over an electronic backyard fence? Because any collaborative filtering system is necessarily premised on having and wanting diverse views." The problem, as Robert Putnam once phrased it, "is that bridging social capital is harder to create than bonding social capital—after all, birds of a feather flock together."

I Really Do Trust My Feed

BJ Fogg of Stanford University's Persuasive Technology Lab began his academic career researching Web credibility and computers as persuasive technology. Of late, his interest has expanded to include emerging media, which he is certain are revolutionizing information filtering and delivery. "Previous theories about social networks are wrong," Fogg states forthrightly. "Because earlier researchers don't get what is happening online."

Fogg says that unlike face-to-face, offline social networks, online networks lend themselves to the easy and friction-free formation of groups. The looser, more extensive social ties that result then lead to more diverse, and ultimately more trustworthy and credible, news and information delivery. "More and more we will be looking at our Facebook feed to see what friends have posted," says Fogg. "That will be how we queue up what is important and credible—and I will do the same for my friends.

"I'm not sure that 'regular news' *ever* brought me greater diversity and credibility than my feed does now," says Fogg. "But clearly in the future, more and more information will be socially filtered in some way. Right now the way that happens is through the feed. But that could change—some current research shows, for example, that the use of online video is super-persuasive, so maybe in the future people will simply go directly to YouTube to be persuaded. Maybe they're doing that in the present as well!"

Fogg's credibility-and-persuasion prescription is not

just some futuristic fantasy, he says. "That's how I live my life now. I really do trust my feed, coming as it does through smart, interesting friends. Otherwise I wouldn't look at the content there—just as I never look at local television news now."

10.

Public Displays of Connection

Long-held conventional beliefs about social networks and the formation of social capital are now running headlong into an area of computer science known as "network theory." The game-changing power of *online* social networks is derived most notably from their facilitation of the formation of groups, thus making it easier than ever before to stay in touch with more people with disparate points of view. By greatly decreasing what academic researchers call the "transactional cost of creating bridging social capital," the tools and technology offered by emerging media enable the finding and sharing of credible news and information through trusted friends and followers—curators and influencers—thus presenting an intriguing possible solution to our ongoing trust dilemma.

Nicco Mele, who ran online operations during the groundbreaking Howard Dean presidential campaign of 2004 (see Chapter 11), describes the theory's three basic laws as Moore's Law, which holds that processing power doubles every two years; Metcalfe's Law, which says that the value of a network depends on number of users of a system; and finally, Reed's Law that the value of a network is directly related to the ability to form groups within it. In sum, Mele says, network theory dictates that "any relatively large group-forming network will inevitably create what is known as the 'network effect,'" the phenomenon whereby a service becomes more valuable as more people use it, thereby encouraging ever-increasing numbers of adopters. "When social capital and community meet online, the result can be a large, group-forming social network that is extremely diverse, highly credible and very powerful," he explains.

Mele echoes Stanford's BJ Fogg in describing his Facebook feed as "a personalized newspaper put out by my friends." He believes "Community, trust and persuasion are the keys to both media and political activism in the future," and adds, "Persuasion and trust are still largely not understood, but there are trends that we *do* understand. The first is that we can always depend on a proliferation of emerging media forms over time; another is that a convergence of communication and community is rapidly approaching."

As we have seen, Mele's belief in the utility of social media's tools and technology is supported by a growing corpus of academic research. When a team at Michigan State University, for example, examined the use of Facebook by undergraduate students over three years, using surveys, interviews, and automated capture of the MSU Facebook site in an effort to understand how and why the students were using the social network, "What we found surprised us," Assistant professor of Telecommunication, Information Studies, and Media Nicole Ellison told the authors of the *New York Times' Freakonomics* blog. "Our survey included questions designed to assess students' 'social capital,' a concept that describes the benefits individuals receive from their relationships with others. Undergraduates who used Facebook intensively had higher bridging social capital scores than those who didn't."

The students found that Facebook helped them maintain or strengthen their relationships with people they didn't know that well, but who still could provide them with useful information and ideas. They used the site to look up old high school acquaintances, to find out information about people in their classes or dorms that might be used to strike up a conversation, to get contact information for friends, and many other activities. Such tools, which enabled them to engage in online self-presentation and connect with others, "will be increasingly part of our social and professional landscape, as social network sites continue to be embraced by businesses, non-profits, civic groups, and political organizations that value the connections these tools support," says Ellison.

In other work undertaken with fellow researcher Kelly Garrett, Ellison's colleague Cliff Lampe has shown that people who receive online information through social networks are more able to articulate opposing viewpoints. Lampe and Garrett's research seems to indicate that sites like

179

Facebook and other social networks, such as Digg.com or Slashdot.org, ironically may function better as information filters than more specialized "social news networks" such as NewsTrust.net, a non-profit community site created in 2005 by former Apple executive Fabrice Florin in a conscious attempt to "help people find and share quality journalism." Since sites like Digg and especially Slashdot are not viewed by their core communities as primarily "political" or "news" sites, the connections established on them through a shared interest in other, less-charged issues such as technology may make them seem a "safe third place—not home, not work," says Lampe. Thus it becomes, perhaps paradoxically, easier (or "safer") to share information and perspectives about politics or news on them, precisely because those networks are not intrinsically political or journalistic in nature.

"You are more likely to hear something you disagree with on Slashdot than on a conventional liberal or conservative single position site," explains Garrett. "There are more sources there that represent multiple viewpoints." Similarly, Facebook, with its stated emphasis on personality and relationships ("Facebook helps you connect and share with the people in your life") can also be regarded as a safe place for the transmission of news and information between trusted friends, even if they disagree vehemently on the topic being shared.

Lampe and Garrett have extensively studied the News-Trust site and suggest that it "may be the wrong model" precisely *because* its visitors are people who deeply care about news, information, journalism, and politics. Hence members may come to the site intent on pushing a particular viewpoint—often political and partisan—and consciously or subconsciously may actually be closed to differing perspectives and analysis. Some may even go so far as to attempt to game the system in some manner in order to further the advocacy of their own ideas and beliefs.

Cast in academic terms, NewsTrust and similar news-oriented social networks may in fact suffer because their community members actually have tight, close connections—bonding social capital ties—rather than the looser, more extensive bridging ties at sites such as Slashdot, Digg, and Facebook, which collect people who have broader interests, such as technology or personal relations, and who then sometimes share news and information about politics and other more tendentious topics.

The Big Picture

Judith Donath, associate professor at M.I.T.'s Media Lab and a faculty fellow at Harvard Law School's Berkman Center for Internet & Society, adds that in the "big picture," social networking technologies will "support and enable a new model of social life, in which people's social circles will consist of many more, but weaker, ties. Though we will continue to have some strong ties (i.e., family and close friends), demographic changes . . . are diminishing the role of social ties in everyday life. Weak ties (e.g., casual acquaintances, colleagues) may not be reliable for long-term support; their strength instead is in providing a wide range of perspectives, information, and opportunities."

Donath's associate Danah Boyd, also a fellow at Harvard's Berkman Center for Internet and Society, says, "Social media (including social network sites, blog tools, mobile technologies, etc.) offer mechanisms by which people can communicate, share information, and hang out . . . social media provides a venue to build and maintain always-on intimate communities." And Donath believes, "As society becomes increasingly dynamic, with access to information playing a growing role, having many diverse connections will be key. Social networking technologies provide people with a low cost (in terms of time and effort) way of making and keeping social connections, enabling a social scenario in which people have huge numbers of diverse, but not very close, acquaintances.

"Does this make us better as a society? Perhaps not," Donath admits. "We can imagine this being a selfish and media-driven world in which everyone vies for attention and no one takes responsibility for one another." But she posits as well that "we can also imagine this being a world in which people are far more accepting of diverse ways and beliefs, one in which people are willing to embrace the new and different."

In "Public Displays of Connection," a 2004 paper they co-authored, Donath and Boyd noted, "In today's society, access to information is a key element of status and power and communication is instant, ubiquitous and mobile. Social networking sites . . . are a product of this emerging culture." They explain that the public display of connections on such sites is a signal that helps others in your network judge your reliability and trustworthiness. New communication technology encourages us to "bridge disparate clusters," which in turn provide us "with access to new knowledge." Trading our previous, offline privacy for shared online public displays of connections enables others to determine our credibility—and by extension, that of the news and information we may then share through the network.

Emerging social media also make it less costly to maintain looser or weak social ties. Such ties, "the kinds that exist among people one knows in a specific and limited context," are "good sources of novel information," say Donath and Boyd. "A person who has many weak yet heterogeneous ties has access to a wide range of information." At a time when Mark Zuckerberg was still formulating Facebook in his dorm room, they predicted that, "In the future, the number of weak ties one can form and maintain may be able to increase substantially, because the type of communication that can be done more cheaply and easily with new technology is well suited for these ties. If this is true, it implies that the technologies that expand one's social network will primarily result in an increase in available information and opportunities—all benefits of a large, heterogeneous network."

In other words, by virtue of being in such a network, where one's identity, trustworthiness, and reliability can be readily assessed, people may access more credible information as well. Although seemingly obvious now, their 2004 vision of "a scenario in which social networking software plays an increasingly important role in our lives," was clearly prescient.

Some leading academic researchers, including Harvard's Robert Putnam, Tom Sander, and Cass Sunstein, continue to question whether social media's filtering function can really be adapted to help solve our credibility deficit and trust dilemma. But many others now recognize and accept the early but suggestive signs of the emerging media's filtering capabilities. Zealots and skeptics alike, however, agree with Putnam, who notes that "although we are a decade into the Internet and eight years into online social networking," more research into the delicate interplay between trust and persuasion, and how they actually function in an online environment, is still sorely needed.

Politics 2.0

On August 29, 2008, just prior to the Republican National Convention in St. Paul, Minnesota, presidential candidate John McCain announced he had chosen Sarah Palin, the governor of Alaska, as his running mate. The surprising choice of the then little-known Palin captured the nation's attention; her status as just the second woman ever to run on a major party ticket was but one among many reasons. Interest in America's already long and hotly contested electoral campaign soon began to reach a fevered pitch.

Several days later, a trusted friend sent me some news about Palin via Facebook. The characteristically brief message—"Check this out!"—referred to the forwarded text of an email from a woman neither of us knew. Her name was Anne Kilkenny and she resided in the small Alaskan city of Wasilla. Kilkenny's email concerned a woman she knew well—Wasilla's former mayor Sarah Palin. "Dear friends," Kilkenny's email began. "So many people have asked me about what I know about Sarah Palin in the last 2 days that I decided to write something up."

A homemaker and regular attendee at Wasilla City Council meetings, Kilkenny had witnessed much of Palin's meteoric political rise at first-hand. She wrote in considerable detail about Palin's record during her six years as Wasilla's mayor, and included a reasonably balanced "CLAIM VS. FACT" assessment ("gutsy: absolutely!") of her personality and politics. Kilkenny's sharp, informative 2400 word missive was meant just for her friends, but as the *Los Angeles Times* reported a month later, "More than 13,700 e-mail responses and half a million Google hits changed all that."

Kilkenny had told her friends to feel free to pass her email along—and so they did, sending it to their friends and followers, who in turn redistributed it in a variety of ways, including via blogs, websites, and social media such as Facebook and Twitter. Moving at the speed of light, the now viral contents of Kilkenny's email soon landed on my computer desktop and on that of millions of others all over the globe. "Who is Sarah Palin?" the world wanted to know—and thus, "Who is Anne Kilkenny?" Was she, and the information in her email, at all credible and worthy of our trust?

A week after receiving the Kilkenny email, I moved back to Cambridge to begin my Shorenstein fellowship. Not long afterwards, another bit of news about Sarah Palin landed in my inbox. Forwarded by a different friend, this email supplied a "list of books Palin tried to have banned" from the local library during her tenure as mayor of Wasilla. The information, if true, had the potential to derail Palin's vice-presidential candidacy almost before it began. But was it?

I was immediately suspicious of the claims made in this second email, which detailed the governor's supposed penchant for book banning, because I trusted neither the sender nor the story. The friend who had passed it on was well known in our social circle as a shoot-from-the-lip liberal prone to exaggeration. In addition, the information he passed on seemed rather suspect. I decided to vet it myself before passing it on.

Sure enough, it turned out to be false; Palin hadn't really banned any books at all. In fact, several works on the list hadn't even been published at the time of their supposed banning, something my credulous friend typically hadn't bothered to check. Soon Internet researchers revealed the entire list simply to be a readily available online compilation of all the "Books Banned at One Time or Another in the United States."

The Anne Kilkenny email about Sarah Palin, on the other hand, proved to be quite credible and indeed very useful. It provided legitimate information not readily available elsewhere at the time—certainly not from the many reports created by the thousands of journalists gathered in St. Paul to cover Palin's impending nomination at the Republican Convention. Ironically, Kilkenny had even delivered the real story about Palin and those supposedly banned library books:

> While Sarah was Mayor of Wasilla she tried to fire our highly respected City Librarian because the

Librarian refused to consider removing from the library some books that Sarah wanted removed. City residents rallied to the defense of the City Librarian and against Palin's attempt at out-and-out censorship, so Palin backed down and withdrew her termination letter.

Like the rest of the news in Anne Kilkenny's email, her information about Wasilla's library books turned out to be very reliable—just as I had presumed it would be, since a trusted friend had sent it to me by via an online social network. Each of the two messages I had received about Sarah Palin had become an instant Internet sensation, rapidly replicated, exponentially amplified and soon propelled into an ongoing national conversation. Each purported to deliver news and information vital to my ability to make an important and informed judgment about people and events that would profoundly affect my life. Each was sent by friends seeking to inform me. One was credible and trustworthy; the other was not. . . .

Campaign and Candidate Media

I had expected that the Shorenstein fellowship would present an opportunity to examine the crisis of confidence in media. But as the 2008 national election campaign continued to unfold, it became obvious that the trust-and-credibility crisis was much broader than that. With political campaigns now beginning to produce and distribute their own media, my questions about the reliability of media reports *about* the candidates were soon joined by questions about the reliability of reports produced *by* the candidates and the campaigns.

Following the announcement that John McCain had chosen Palin as his running mate, public interest in America's 2008 national election soared. So too did voter, candidate and party reliance on new media tools—especially those that facilitated controlling one's own media. The race for the presidency, with its viral emails, social networks, user-generated videos, fact-checking websites, MySpace and YouTube debates and other online innovations, provided another ideal prism through which to examine the rise of social media and to assess both their utility as trust filters and impact on longstanding media and political brands alike.

The political importance of the emerging media was most apparent in the successful candidacy of Barack Obama, whose online-focused campaign revolutionized modern politics in ways that are still coming into focus. No previous candidate or campaign had ever adopted technology and the Internet as the heart of its operation or used it on such a scale. Aided by MyBarackObama.com, a Facebook-like social network created with the assistance of that company's co-founder Chris Hughes, and employing a team of young, Web-savvy programmers and developers who had cut their teeth on Howard Dean's 2004 presidential primary campaign, the underdog Obama embraced social media and hugged them closely all the way to the White House. He used social networks to raise record-breaking amounts of money—more than $500 million from 3 million donors who made a total of 6.5 million donations online. He also used the new media to circumvent longstanding media and political brands by communicating with his supporters directly and interactively.

"On MyBarackObama.com, or MyBO, Obama's own socnet, 2 million profiles were created," Jose Antonio Vargas noted in one *Washington Post* campaign post-mortem. "In addition, 200,000 offline events were planned, about 400,000 blog posts were written and more than 35,000 volunteer groups were created. . . . Some 3 million calls were made in the final four days of the campaign using MyBO's virtual phone-banking platform.

"Obama has 5 million supporters in other socnets." Vargas reported. "He maintained a profile in more than 15 online communities, including BlackPlanet, a MySpace for African Americans, and Eons, a Facebook for baby boomers. On Facebook, where about 3.2 million signed up as his supporters, a group called Students for Barack Obama was created in July 2007. It was so effective at energizing college-age voters that senior aides made it an official part of the campaign the following spring."

For the first time, many voters also used that same emerging media, its powerful tools and looser, more extensive social networks to communicate directly with their peers about the election. Media platforms that hadn't even existed in the previous presidential election cycle just four years earlier began to play crucial roles in political campaigns and the delivery of information about them. YouTube, as we have seen, essentially deflated the presidential hopes of Virginia Senator George Allen when it captured his "Macaca" moment. Other non-professional videos later uploaded to YouTube also went viral and had great impact on the campaigns, gaining such popularity that they were picked up and reported on by the legacy media. (Prominent examples include a mash-up of Republican presidential candidate Mitt Romney declaring his support for abortion and gay rights—positions he later renounced—and a spoof of the famous Apple Super Bowl Ad that compared Senator Hillary Clinton to the oppressive system described in George Orwell's *1984*.)

By spring 2008, YouTube had launched its Citizen News channel to highlight user-generated news. Major legacy media organizations such as CNN and ABC increasingly reached out to the company, along with such other social media platforms as MySpace and Facebook, seeking access to their growing audiences. MySpace partnered with the Commission on Presidential Debates to create an online town hall forum where citizens could discuss a debate, submit questions, watch it televised live and then take issue quizzes, while the moderator, CBS correspondent Bob Schieffer often cited the social network during the broadcast. As Kathleen Hall Jamieson, a professor at the University of Pennsylvania and a chronicler of presidential races for more than forty years, told the *Washington Post*, "In the past there was only a passive relationship between the producer and the audience. But the audience has also become the producer. That's very empowering—and a huge change.

"There's a dark side to this, of course," Jamieson continued. "Voters can only read and watch and interact with everything they agree with, creating a hyper-partisan and largely uninformed electorate. But there's also a bright side where an informed and engaged electorate can participate in discussions that are relevant to the political process. Which way we'll eventually go, we'll have to see."

The Senior Fellow at the Institute of Nonexistence

As the campaign unfolded, making credibility assessments of political news and information quickly became a cottage industry of sorts, as issues of persuasion and trust continually arose involving both the candidates and the media. The repeated use of untrue information and false claims by Senator McCain and Governor Palin—and repeated media corrections of them—became a constant campaign drumbeat. Palin's persistent claims about her alleged opposition to Alaska's so-called Bridge to Nowhere, for example, and journalists' reports that she actually had supported it, created ongoing friction between the media and the Republicans. They also moved Democrat Barack Obama to remark, "You can't just make stuff up. You can't just recreate yourself, you can't just reinvent yourself. The American people aren't stupid."

Various news organizations created websites specifically to check the veracity of claims made by the candidates and their campaigns. The *St. Petersburg Times* and *Congressional*

Quarterly created PolitiFact.com, the Washingtonpost.com began Fact-Checker, which often looked at third-party and campaign ads circulated via YouTube and awarded "Pinocchios" to candidates who bent the truth, and the George Polk Prize-winning blog *Talking Points Memo* turned its "Veracifier" page into one of the most-trafficked News & Politics pages of YouTube. Soon sites like PolitiFact.com and Factcheck.org were drawn directly into the fray, even finding their evaluations used—and misused—in campaign ads, such as one released by the McCain-Palin campaign, which Factcheck promptly charged "has altered our message in a fashion we consider less than honest." Republican campaign officials in turn claimed that Senator Obama had also made a number of false claims and complained that the media was biased in favor of the Democratic candidate instead of holding him to account.

Relations between the Republicans and the legacy media deteriorated drastically as the McCain/Palin camp charged that "advocacy" on behalf of Barack Obama was behind the media's reports of falsehoods and persistent questions about the credibility of claims being made by Republicans. Steve Schmidt, the day-to-day manager of McCain's campaign, blasted the *New York Times* in particular as being "completely, totally 150 percent in the tank for the Democratic candidate." Schmidt added that the *Times* had "cast aside its journalistic integrity to advocate for the defeat of John McCain."

The Republican assault on the media was met with

fierce resistance. *Times* editor Bill Keller responded that his paper was committed to covering the candidates fairly, saying, "It's our job to ask hard questions, fact-check their statements and advertising, examine their programs, positions, biographies and advisors." *Newsweek* columnist and *NBC News* correspondent Jonathan Alter was even tougher. "We're seeing the emergence of a 'smear gap,'" Alter noted. "John McCain making stuff up about Barack Obama, and Obama trying to figure out how hard he should hit back.

"McCain's campaign has been resorting to charges that are patently false," Alter added. "In the past, plainly deceptive ads were the province of the Republican National Committee or the Democratic National Committee or independent committees free to fling mud that didn't bear the fingerprints of candidates. But not this time. These smears come directly from the candidate." While castigating McCain for his "untrue charges and false spots," and urging him to "stop lying about his opponent," Alter also noted, "[O]ne of the wonders of the Web is that it's now possible for neutral observers to determine the truth or falsity of various attacks, and to have that information instantly available to anyone."

Throughout the campaign, legacy media brands also continued to be plagued by their own credibility problems. Shortly after the election, for example, it was revealed that a putative McCain policy adviser named Martin Eisenstadt, who had been quoted by such traditional media sources as NBC News, the *New Republic*, and the *Los Angeles Times*, didn't actually exist. Although he had been widely cited as the source of leaks from within the Republican campaign claiming that Sarah Palin "did not know that Africa was a continent," it turned out that Eisenstadt was just a figment of the fertile imaginations of a pair of obscure filmmakers who had created him in hopes of selling a television show. "Martin Eisenstadt doesn't exist," the *New York Times* noted. "His blog does, but it's a put-on. The think tank where he is a senior fellow—the Harding Institute for Freedom and Democracy—is just a website. The TV clips of him on You-Tube are fakes." The paper ironically dubbed Eisenstadt the "Senior Fellow at the Institute of Nonexistence."

The filmmakers argued to the *Times* that "the blame lies not with them but with shoddiness" in the media itself. They were right—an NBC News spokesman explained away the network's misreporting by saying someone in its newsroom had received the misinformation about Palin in an email from a colleague and *assumed* it had been checked out. (Sound familiar?) "It had not been vetted," the spokesman admitted. In essence, as the *Times* reported, the media fell for the fake material "despite ample warnings online about Eisenstadt, including the work of one blogger who spent months chasing the illusion around cyberspace, trying to debunk it."

What Wins Today Is Being in Touch
Have we now come full circle, as Jonathan Alter suggests? Have social media transformed America's political

campaigning forever? With the legacy media under assault and reliance on the credibility of their once-trusted brands shattered, have "the wonders of the Web" become the solution to both our media and political trust dilemmas? Online politics pioneer Nicco Mele, who led Howard Dean's Internet team and helped Barack Obama win a Senate seat, and whose associates later ran Obama's presidential online operations, is among the believers.

Mele, whose father was a diplomat, spent part of his childhood abroad. An avid fan of professional baseball in general and of his hometown New York Mets in particular, he was often frustrated by the difficulty of finding timely information about the team—until he discovered the Internet. "As a kid, I had one of the first IP addresses ever registered in Malaysia," he remembers.

In addition to being a sports aficionado and tech wizard, Mele is a self-described political junkie. Back in the United States years later and with little money after having graduated from college, he searched for an inexpensive way to find trusted and timely political information. Once again the Internet provided a solution. "I became a devotee of Hotline, the political insider's journal, when I could access it for free while in college," he says. "After that, I couldn't afford it— but I found that some of the early blogs about politics could deliver some of the same experience for free. Also, the first meetups ever organized online began around then, and that was powerful for me as well."

In early 2003 Mele was invited by a friend to a meetup

of people interested in supporting Dean, then the obscure governor of Vermont who was considering a run for the presidency. Soon Mele volunteered his technical skills to advance Dean's cause through online activism.

"In political terms, online was a backwater until 2003 and Dean's campaign," Mele recalls. "We were inspired by the success of groups like MoveOn.org—they are owed an enormous debt that is often overlooked—and we took their best practices and applied them for first time to a candidate instead of to issues, as MoveOn had done."

Mele joined Dean's campaign effort in late April 2003, when few major media outlets had ever even mentioned Dean's name. "He was not on their radar at all," Mele remembers. "So in lieu of the mainstream media, we began to use the political bloggers on the left, like My DD and the Daily Kos."

The year 2003 marked the ascent of the so-called "blogosphere" as a new force in media and politics. "Remember, there was no YouTube then, no Facebook," says Mele. "Blogs were the world." Politically progressive bloggers, along with the grassroots antiwar coalition opposing the coming conflict in Iraq, soon became "key differentiators for Dean," Mele says. The Dean campaign started to gain supporters without even being noticed by legacy media. As a result, one of his early speeches went viral on the Internet, and even though the once all-important *New York Times* had hardly mentioned his name, "random people on the Net began making ring tones from the speech!" Mele recalls.

His colleague Jim Brayton was also new to electoral

politics; the Dean campaign was his first involvement with any campaign. "I was an Internet administrator for a small business and a blogger, but actually more into technical stuff than content," Brayton says. "I saw that Howard Dean had started blogging too. It interested me to see that a political candidate was blogging, so I offered to help via email."

Brayton's email went unheeded, however. Frustrated, he decided just to show up at Dean's headquarters. "I went down in February 2003 and offered tech help, and they accepted. The first thing I did was to set up an 'auto-responder' so that no other email would go unanswered," he says. "I did that for few weeks, then they brought in Nicco and we began to work together."

The Dean campaign was groundbreaking in its effort to make extensive use of the Internet, although Mele is quick to credit others before him. As examples, he points to Bill Bradley's success in persuading the Federal Elections Commission in 1999 to match funds raised via online credit card use and John McCain then raising millions of dollars over the Internet the following year in the week after the New Hampshire primary. Bradley, Mele remembers, even posted a prescient statement on his campaign website, saying, "I don't know if it's going to be 2000 or 2004, but one presidential election very soon is going to be decided because somebody understood how important the Internet was, respected the people who were on it, and wanted them to be a part of something new and different that made America a better place."

Both Mele and Brayton describe the Dean effort as revolutionary not only in terms of fundraising but also for its use of email as a way of communicating directly with voters. "The dynamic of the campaign created some unusual technical needs," says Mele. "Dean's was first in terms of the application of many of these ideas."

"Then the software got better," Brayton adds, citing the Blogger technology developed by Evan Williams (later of Twitter fame) as an example. "The average person could finally handle it."

Using blogs as a means of attracting attention obviously didn't fit into any historical or traditional methods of political campaigning. "It was a default position, in that we had little choice, but it was also a conscious decision," says Brayton. "We just wanted to get anyone we could talking or writing about him. When you're a no-name candidate, anything that works, you do more of—and in Dean's case, that meant blogs and meetups. For us to get covered by the Daily Kos blog was exciting."

By May 2003, all the online activity, coupled with Dean's performance in the first Democratic presidential primary debate, had combined to make him a grassroots celebrity. The next month, his surprising success in online fundraising "changed the equation totally," Mele recalls.

"Viewed in retrospect, it's obvious that fundraising was the primary online app," says Brayton. "Being a political novice, it all seemed natural to me. So only after June filing could I see this was really something different."

"Suddenly Howard Dean had raised 15 million dollars," Mele adds. "More than anyone else in the race—and it happened online while the political establishment wasn't even watching!" As Brayton concludes, "The money we raised online gave us viability."

Both Brayton and Mele see a direct line from Howard Dean to Barack Obama, and both worked for Obama during his 2006 campaign for the Senate. "The Obama campaign called us and said they wanted to build an online operation similar to the Dean thing," says Mele. "In a way it was more challenging though, because with Dean the fire had started and we just threw logs on it."

On the other hand, the Dean campaign had paved the way in many respects. "Email testing, social networking, using credit cards to pay online. . . . All those ideas had already been tested when Obama came," says Brayton.

In the fall of 2006, Facebook stopped being exclusive to academia and opened to the world. "Obama was interested in it right from the start," says Brayton, who spoke to Facebook co-founder Chris Hughes early on while looking for suggestions as to how to emulate it. Then Obama decided to run for president. "The main takeaway from the 2006 Senate campaign was that Obama started building his email list for 2008," says Mele. "And that became arguably his major asset going forward."

The Internet and emerging social media proved crucial to Obama's successful presidential campaign. "Hillary would have won but for Internet," Mele says point blank.

"Social media gave the underdog the power to challenge her, so it was transformative. Online fundraising in particular was an early equalizer. Just having name recognition and access to traditional sources of money was no longer sufficient." Obama later exploited the Net in other important ways, such as field campaigning and staff organizing.

Both Mele and Brayton say adept use of the Internet and social media is no guarantee of political success, however, and they offer various caveats for Net enthusiasts. "Number one, although it gave Obama the ability to raise money, it did the same for Dean—but he didn't win," notes Mele. "Number two, let's not forget that while Obama raised money on the Net, he spent most of it on television—still a dominant political medium. And number three, we should also remember that the Internet hasn't yet significantly impacted any other race."

The Other Side of the Aisle

Mindy Finn is another pioneer of online politics, albeit on the other side of the American political aisle from Brayton and Mele. Finn and fellow advocate Patrick Ruffini are partners at Engage, a political strategy and communications agency. As her corporate biography states, "Finn jumped on the Internet bandwagon . . . well before online politics was cool." She has held senior new media positions in both the Bush/Cheney presidential campaign in 2004 and that of Mitt Romney in 2008. As Director of "e" Strategy for the Romney campaign, she used web video, social networking, blog outreach, user-generated content gathering, email list building, and online advertising to communicate the candidate's message, raise money, and mobilize a base of support.

Finn disagrees with the notion that Republicans have lagged behind Democrats in their understanding and use of the Internet and social media. "The conventional wisdom is that the Democrats were ahead in 2008, but we push back on that," Finn said in a 2011 interview. "The right was not behind leading up to 2008. Yes, there was some disparity in the adoption of the tools, but it wasn't the tools that drove the election. The key drivers were message and narrative. The political momentum in 2008 was all on the Democrats' side. They then used the tools well and were able to capture that dynamic." After Obama's victory, Finn believes, Republican politicians understood and employed the power of online communications more so than anyone on left. "We have now seized the new tools like Facebook and Twitter," she says.

Finn echoes Nicco Mele's observation that although "MoveOn and the Dean campaign no doubt set the pace, in actuality Dean's use of the Internet didn't get him elected— or anyone else for that matter—and was mainly used to raise funds." As a result, she says, "People could partly dismiss the power of online because Dean had no electoral victory."

But that was then. "In any event, social media is now absolutely central to all political campaigns," Finn avers. "The low barrier to entry gets you buzz, name recognition and effective money raising, all at a low, low cost."

She points out that in the 2010 mid-term elections,

many Republicans surpassed Democrats in their adroit use of social media and touts three Republicans—Marco Rubio in Florida, Sean Duffy in Wisconsin, and Rick Perry in Texas—for running the best online campaigns of 2010.

"Marco Rubio got into the Senate race eighteen months before the election; if you looked at polls or money raising then, it seemed clear he had no shot," she says. "At least that was conventional wisdom, but those with their finger on the online pulse knew better. There was lots of blog action and activity on the social networks, along with fundraising. Rubio was constantly reaching out through social media in ways [his primary opponent Governor Charlie] Crist wasn't. He had video on his website, did lots of blogger calls, and so on. Just as with Obama's campaign against Hillary, without that very early focus online, Rubio never would have been elected."

Similarly, Sean Duffy in Wisconsin was also thought not to have a shot against a powerful incumbent, longtime Democratic officeholder David Obey. "Right from the beginning social media was very important in Duffy's campaign," says Finn. "He used flip cam videos very effectively; he was constantly on Facebook to build an audience. . . . The people who supported him then became the distributors, and shared because they *cared*." As a result, she says, Duffy built so much excitement online it ultimately drove Obey out of the race.

Finally, Finn cites Texas Governor Rick Perry's campaign. "You would think as an incumbent he would be the establishment candidate, but he had a primary challenge and was behind in the race," she points out. "So Perry immediately changed his tone and message to become more like the Tea Party and began to run a totally different campaign. There was no direct mail, no robo calls, and not a lot of television. Instead it was an online-driven candidacy and it dramatically shifted public perception of him. It also helped that Perry personally is an active tweeter and sometimes spent an entire day with top bloggers. These days, that's worth an entire day with top funders—that's how tangible social media is now."

Like her Democratic counterparts, Finn is certain there's a transformational shift going on. "Social networking is now the very foundation of your campaign, it supports everything you do," she says. "It can't be compared to other media and you just can't run old media campaigns like in the past."

The candidates' newfound reliance on emerging media and the Internet is mirrored by that of the voters, as a recent Pew study shows. Nearly three-quarters of all adult Internet users (representing more than half of all adults in the U.S.) went online to get news or information about the elections in 2010, or to get involved in the campaign in one way or another. Many simply sought political news; others wanted to take part in specific political activities, such as watching videos, sharing, and fact checking political claims. More than one in five used Twitter or a social networking site. Predictably, however, they were still of two minds as to the Internet's efficacy. As the study's authors noted in an overview, "They hold mixed views about the impact of the Internet. . . . It provides diverse sources, but makes it harder to find truthful sources."

The trend is quite clear; the 2010 figures on social media and Internet use for obtaining political information and fostering engagement far surpass the 2006 mid-term figures as well as those of 2008 national election, and there is no end in sight. In the forthcoming election and beyond, these figures will no doubt continue to rise dramatically.

Mindy Finn forecasts that, "Both Republicans and Democrats will put an enormous effort into this. Social networks are already having a major impact on all political calculations now taking place. They even are influencing the types of candidates we get, since they create more of an open door to run than in past. People no longer have to feel counted out in advance. Anyone paying attention out there knows this is a game changer."

National candidate Sarah Palin is an interesting case study, says Finn. "Palin became expert at using Facebook and Twitter, and then took it from there to cable TV news and even her reality show," she notes. "She absolutely *gets* this—and her fans are rabid friends and followers in the social media."

Finn's conclusion jibes with that of Mele, Brayton, and other online politics pioneers. "Going forward it will be social networks and mobile in particular that will be huge," she posits. "Everything will be much more distributed, there will be more partnerships, and much more outreach to bloggers and the community. Politicians will be forced to run less insular campaigns than in the past. Just being rich and throwing your weight around is not what wins campaigns anymore—instead it's your networks, organically built. What wins today is being in touch and then responding to in an authentic way. When candidates actually believe this, and participate personally, then they see it power and that gets them to believe—plus they benefit in many ways from the direct feedback. The candidate who can best tap into social will likely win in 2012. This whole new dynamic, in essence, is having a dramatic impact on all political calculations going forward."

12.

The Daily Me vs. The Daily We

With brands of all sorts in tatters, a flood of information inundating us, legacy gatekeeping disrupted, and "quality journalism" in decline—at least as defined by corporate executives such as Bill Keller of the *New York Times* and Google's Eric Schmidt—how can average citizens hope to find truly credible news and information? Moreover, are they even interested in doing so—or will they gravitate instead to what some term "The Daily Me," an online mediascape where each of us is our own editor and gatekeeper? Rather than presenting an unbiased array of useful, trustworthy information and different, unexpected viewpoints, do the Internet and interactive social media instead cause us to become more close-minded? Do they merely reinforce what we already think we know, as various pundits, professors, experts, and executives claim?

"There's pretty good evidence that we generally don't truly want good information—but rather information that confirms our prejudices," says one proponent of this view, *Times* columnist Nicholas Kristof. "We may believe intellectually in the clash of opinions, but in practice we like to embed ourselves in the reassuring womb of an echo chamber. And if that's the trend, God save us from ourselves."

Some prominent academics, including Cass Sunstein of Harvard and Nicholas Negroponte of MIT, also promote this perspective. Sunstein, for example, contends that we are now witnessing an overall decline in the influence of what he terms "general interest intermediaries," and an increase in highly specialized arenas for information such as partisan cable television channels or websites, which allow us to "personalize" the news we receive. In such a culture, he argues, we have the ability to see only what already interests us and we ignore exposure to the varied concerns and opinions of fellow citizens.

The problem of having too much information, Sunstein says, leads inevitably to a nightmare of limitless options. We then respond by carefully filtering out opposing or alternative viewpoints to create an ideologically exclusive "Daily Me," while inevitably gravitating toward media that reinforce our views.

This effect, Sunstein and Kristof argue, tends to insulate us in "our own hermetically sealed political chambers." Worse, the continued decline of traditional news media will only "accelerate the rise of The Daily Me, and we'll be irritated less by what we read and find our wisdom confirmed more often," Kristof has written. The danger, he says, "is that this self-selected 'news' acts as a narcotic, lulling us into a self-confident stupor through which we will perceive in blacks and whites a world that typically unfolds in grays."

As evidence, Kristof cites books such as Bill Bishop's *The Big Sort: Why the Clustering of Like-Minded America is Tearing Us Apart*. In it, Bishop argues that Americans increasingly segregate themselves into communities, clubs, and churches and surround themselves with people who think the way they do. Many live in "landslide counties," largely either Democratic or Republican, which rarely switch political allegiance. "The nation grows more politically segregated—and the benefit that ought to come with having a variety of opinions is lost to the righteousness that is the special entitlement of homogeneous groups," says Bishop.

What Bishop calls "clusters," however, are real-world social networks—the kind Robert Putnam wrote about in *Bowling Alone*—and not online social networks like Facebook. Online networks are different; by exposing us to larger groups of people with more varied information and perspectives, they may tend to drive us more to trust and persuasion than polarization and reinforcement. In fact, many social media researchers now believe the connections forged and maintained in such networks may actually *expand* exposure to conflicting ideas and allow users to engage in dialogue that sometimes results in changed opinions and attitudes. "The claim that people isolate themselves is based on faulty assumptions," says Kelly Garrett of Ohio State University. "The situation is far more complex.

"There is an 'echo chamber' possibility, but it's not

necessarily so," he explains. "And it's certainly not causing the dire consequences seen by Sunstein. There *is* that risk, but as to his claim of a more balkanized society, there's no real evidence that this is so. It's a mistake to claim online news, or the Internet itself, are essentially either good or bad; that's too reductionist an argument. There is mixed evidence on both sides, but ultimately society will probably not become 'cyber-balkanized.'"

"Better filters are needed to create personalized sets of news," says Paul Resnick of University of Michigan. "But will they lead, as per Sunstein, to a world of echo chambers? Our conclusion is no, that is not were we are headed . . . at least not for everyone. Homophily is a fear, sure, but it won't happen automatically.

"That concern comes from over-extrapolation," Resnick notes. "Given a choice between two partisan channels, people will pick the one they agree with, but evidence seems to show that these same people don't want *only* confirmation of their views. In politics, for example, it can be useful to know what the 'other side' thinks. In addition, the social norm says we should be willing to hear the other side. That's a basic democratic impulse and value that is widely shared."

Web credibility researcher Eszter Hargittai agrees. Hargittai was part of a team that studied Web links among the writings of leading conservative and liberal bloggers. "As for Sunstein's hypothesis," she says, "we certainly don't have conclusive information at all that would indicate growing isolation and balkanization over time."

Hargittai's team began the study's abstract by noting, "With the increasing spread of information technologies and their potential to filter content, some have argued that people will abandon the reading of dissenting political opinions in favor of material closely aligned with their own ideological position." In fact, however, the team concluded, "Bloggers across the political spectrum . . . address each others' writings substantively, both in agreement and disagreement." In other words, the fragmenting potential of information technologies—at least as measured by Web links among bloggers—does not support Sunstein's selective exposure argument.

Friends with Friends Or Friends with the Feed

The self-induced personalization of a Daily Me may pale, however, when compared to a newer, more hidden phenomenon: the unseen, machine-created personalization now happening automatically on many websites. As corporations increasingly fine-tune their ability to use our online history as a filter to narrow down what they present to us, will we all come to live in our own unique information universe, receiving only familiar, pleasant news that merely confirms our beliefs?

It's already happening, says Eli Pariser, and "since the filters are invisible, we don't even know what is being hidden from us." Pariser, an online organizer and former director of MoveOn.org, is the author of *The Filter Bubble: What the Internet Is Hiding From You*. He fears the development of a new updated Daily Me, one in which machines and algorithms—not family, friends, and followers—determine what we see, read, and hear online. As corporations increasingly strive to tailor their services, including news and search results, to our perceived individual tastes, an unseen algorithmic personalization may be taking over. With little exposure to information that can challenge us, says Pariser, the custom-made, automated digital personalization of "server farms, secret algorithms, and geeky entrepreneurs" poses an even great danger than the self-created personalization decried by Sunstein.

With Facebook, Google, Yahoo and many other leading websites now personalizing the delivery of news and information to our perceived interests, we are living within what Pariser calls "Internet filter bubbles." As algorithms increasingly provide information based on such factors as our location, search history and past online behavior, he believes we are becoming surrounded by "information junk food," with ideas that may challenge us or teach us something new being filtered out by machines before we even see them.

"Sunstein says reinforcement is sought after by people; I say it's not all about people's choice," Pariser said in an interview. "At least with Sunstein's 'Daily Me,' you create it. But with machine personalization, the lack of diversity is involuntary, so more choices no longer necessarily mean exposure to more streams of information. Instead, a custom-tailored information environment is creeping up on us.

"Do we know how exactly Facebook personalizes?" he asks. "Consumers can't see how it works; we can't see the code. But the result of what Facebook is doing means losing contact with new things and ideas." Facebook's personalized news feed and Google's personalized search, Pariser says, show us only information thought to agree with our interests and predilections and create "the impression that our narrow self-interest is all that exists." The detrimental effects of this algorithmic editing of the web include making us more susceptible to "invisible auto-propaganda that is indoctrinating us with our own ideas and locking us into information silos."

The politically progressive Pariser, for example, became upset when he noticed that Facebook's personalization algorithm was downplaying what more conservative friends posted on his news feed. "I had deliberately reached out to get exposure to a very broad range of information and perspectives," he told me. "And then Facebook filtered them back out of my feed!" Small wonder Pariser argues that the unseen personalization now happening on leading websites presents an even greater threat to finding useful, trustworthy information than that of Sunstein's "Daily Me."

Pariser's thoughts as to how much personalized filtering is actually happening on the Web, and whether it is harmful or not, are not universally shared. Facebook's Chief Operating Officer Sheryl Sandberg, not surprisingly, is far more sanguine about its effects. In fact, Sandberg—whose company launched "Instant Personalization" in April 2010 and later expanded its use in a wide-sweeping 2011 redesign—believes any website that *isn't* tailored to a specific user's interest will soon be an anachronism. Speaking at a 2010 advertising conference, Sandberg told publisher Arianna Huffington, "People don't want something targeted to the whole world," she says. "They want something that reflects what they want to see and know."

Sandberg's vision of the future led Marshall Kirkpatrick of the *ReadWriteWeb* technology blog to caution, "So much for *all the news that's fit to print*—Sandberg's vision of the future sounds more like all the news that's relevant to your taste profile and social graph. Is that emphasis on personalization, which Facebook is better suited to power than any other company in history, a good or bad thing for media and the democracy it ought to fuel?"

Since Facebook's instant personalization had been implemented with little fanfare, most of its users were unaware they were seeing only certain updates from certain friends. Like Pariser, they were being fed invisible homogeneity instead of the diversity they might have been seeking. Many didn't realize until long after the fact that they could reset the default controls on their feed to show greater variety than the new automatic setting allowed—in effect re-personalizing and customizing their feeds.

Shortly before Facebook's Instant Personalization, Google launched something it called "Personalized Search for everyone." As a result, people entering the same search terms began to get different, customized, and "personalized" results. "Now when you search using Google," a company blog explained, "we will be able to better provide you with the most relevant results possible." Rather than improving the user experience, Pariser sees such automated personalization as an unfortunate consequence of corporate power in the digital age. "There is no standard Google anymore," he claims. "Instead, the company uses 57 different things it knows about its users to serve customized search results."

Harvard law professor Jonathan Zittrain, co-director of Harvard's Berkman Center for Internet and Society, disputes the extent to which Google's personalization and filtering distort search results. Zittrain told *Slate* magazine's Jacob Weisberg that Google isn't really doing what Pariser says it is, and that "the effects of search personalization have been light."

Some say the personalization offered by machines might even be positive and could one day lead to a "Daily We." As Marshall Kirkpatrick observed, "If the Facebook vision of 'instant personalization' comes true . . . you'll be shown first and foremost content on topics that you have expressed an interest in already, which is described in the same ways you describe your interests and that is deemed valid by people you trust." Such an approach might result not in polarization but instead in a newfound ability to "deep dive into specialized news and analysis, on the topics that are most important to you," says Kirkpatrick, making them "easier to discover than ever before," as well as offering "a new level of subject-level sophistication, detail and efficiency" to a wider variety of people.

Will algorithmic personalization lead to a Daily Me or

a Daily We? The jury is still out, but as Kirkpatrick concluded, "Now is a good time to consider these questions, because the era of *all the news the algorithm calculates you'll like* is very fast approaching." Many major news brands are already moving rapidly to create personalized information engines that will tailor search results to ones which users are likely to agree with. As Jacob Weisberg noted in *Slate*, the *Washington Post* launched Trove, "a personalized news and information engine," the *New York Times* created News. me, a subscription-based "personalized news service," and "services like Flipboard and Zite, which create personalized 'magazines' for tablets based on your Facebook and Twitter feeds, are new Silicon Valley darlings."

Even if the results of algorithm-driven news are positive, however, we still face a dilemma: With huge amounts of information available at a click, how can we tell what is trustworthy and what is not? Will we even make the effort? "We may *want* to do the right things to find credible news . . . but will we? Can we*?"* asks Miriam Metzger of the University of California at Santa Barbara. "People know they 'should' critically analyze the information they obtain online, yet they rarely have the time or energy to do it."

Truth, Reframed

MacArthur Foundation "genius" award winner Howard Gardner shares Metzger's doubts. Gardner, a professor of cognition and education, as well as of psychology, at Harvard University, is the author of more than two dozen books. His most recent, *Truth, Beauty and Goodness Reframed* examines the related questions of how we define and how we can find truth in the face of what he calls "the ever expanding, ever more powerful" digital media. "The advent of digital media has not fundamentally altered the establishment of truth," Gardner writes. "But any expert who wants to remain current, or even relevant, must rethink his or her processes in light of the digital media—what they emphasize, what they afford, and what they may render obscure or invisible."

Echoing Eric Schmidt, Gardner wonders, "If we consider the welter of information and misinformation available on any search engine, how can we possibly determine what is true, or even whether the search for the truth has become a fool's errand?" The new media, he writes, "have ushered in a chaotic state of affairs, a mélange of claims and counterclaims; an unparalleled mixture of creations, constantly being revised; and an ethical landscape that is unregulated, confusing, indeed largely unexamined." Faced with such disorder, how can we determine what is true, "When we can all present ourselves on social network sites anyway we want?" he asks. "Or when blogs can claim without evidence or consequence that the current American president was born in Kenya?"

Gardner amplified his book's thesis in a 2011 interview. "Do people really want to know the truth in the digital era? The proliferation of information opens the possibility of really getting close to it—but simultaneously makes it easy to ignore!" he said, adding that he worries the young, in particular, are interested "principally in authenticity and transparency, more than sheer 'truthfulness.'" His research, Gardner told me, shows that "many young people are interested in precisely what you are asking, but they are afraid to make false steps or to take a chance." Instead, he says, "They use arbiters of truth— Jon Stewart, Stephen Colbert—and rely on them to 'tell me what I need to know.' They have a reluctance to go it alone, to take risks, to find out for themselves . . . and thus less interest in truth and less concern or caring about 'getting it right.'

"In principle we can do it—we *can* find the truth," says Gardner. "Truth, it must be made clear, is not a question of bias or gut instinct; it consists of carefully-arrived-at conclusions on the basis of cool and consistent review of the evidence. In fact, if we have the patience to find it, and are willing to work hard, we are now in stronger position than ever to figure it out.

"Whether the young approach these web 2.0 opportunities—wikis, blogs, and tweets—with a focus on truthfulness, with complete indifference to truth value, or somewhere in the 'truthiness' between, will have significant consequences," he concludes. "Can they live with more plurality and fragmentation—or do they just not care?"

Although academics such as Gardner, Garrett, Metzger, and others differ from Sunstein et al in their assessments of emerging media's role as a credibility filter, all agree that the available data and research is far from definitive. As Cliff Lampe of Michigan State University notes, "This is an amazingly rich field where almost nothing has been done." Miriam Metzger, whose work centers around media, information technology and trust, adds, "It's a really interesting question, though there are not a lot of people publishing about this yet—at least in traditional academic outlets." As she explains, "The early thought on social networks was that people would self-select—so-called 'homophily'—but evidence now suggests that in seeking political information, it works the other way around. . . . We don't really have an answer yet—but what you suggest seems right. It certainly passes the smell test."

Still the devil, Metzger points out, is in the details. "In some cases, social networks may lead to polarization—but in others, they may not," she says. "The key questions are: What are the bonding conditions? What precisely are the bridging conditions? What precise kind of networks actually facilitate, in terms of credibility? On sites like Facebook and Slashdot, the communities already have basic relations of trust. You can find 'people like me' there, a sense of real community and trust. So in both cases the baseline level of trust is there first; thus people already feel secure in the environment."

In sum, "There is not much data yet, and that which exists is preliminary and mixed," cautions Metzger. "But my feeling is that social networks as filters are very much happening already." Danah Boyd, Senior Researcher at Microsoft Research, adds more forthrightly, "Of course they are being used as filters—but how, why, and at what costs are the research questions."

"How sophisticated are people about this?" asks Metzger. "Very, it turns out. They are already using media tools as trust filters in a sophisticated way." She says the burden of credibility evaluation has increased because of technology and the consequent flood of too much information, but she agrees with Howard Gardner that technology paradoxically also helps with the problem.

"It's a double-edged sword," Metzger explains. "Technology changes the problem, making it more urgent and giving us a greater burden to verify, but it also provides new tools to grapple with credibility questions. This means the technology is opening up more possibilities for solutions—as well as simultaneously contributing to the problem. So the problem changes with each particular situation."

Some researchers looking into automated personalization and recommender systems believe the danger is only temporary. Dean Eckles is among them. A social-cognitive scientist and doctoral candidate at Stanford University, where he worked at BJ Fogg's Persuasive Technology Lab, Eckles has also been a researcher at Facebook, Nokia, and Yahoo. His research centers around social interactions through communications technologies and the "social relationships" we have with the technologies themselves.

Eckles says that filter bubbles such as Pariser fears are a real concern—for now. "Simple filtering mechanisms *are* more likely to push agreement and homophily," says Eckles. "So yes, if I am now embedded in a liberal social network, or with lots of liberal friends, I will be more likely to see just their content." But the learning machines are adaptive, "so they will get better over time," Eckles assures. "The problem is that most people not yet even aware of personalization filtering by Facebook and others, so they tend to see it as magic, instead of something they can control.

"The challenge is to make process intelligible to the user," says Eckles. "Is the feed itself a distinct social actor—or a transparent tool? In other words, are you friends with friends—or friends with the feed? We have to move to more active curation, where you specify the filters you use. You must be an active curator of your own information feed so it can be controlled."

Clearly, much more research and empirical evidence is still needed. One reason is that relatively few studies have yet to look at what users actually do to assess online credibility. As Eszter Hargittai says, "While considerable prior research has examined what users *claim* to do in order to find credible information online, research that compares actual and reported behavior is less common."

Despite the many lingering questions, Stanford University's BJ Fogg stands convinced. "Social networks are the Big New Channel," asserts Fogg, "People don't go to Facebook to 'get something done' but to browse in an interruptible and seducible way. When you find news there, you feel like it's a discovery—not something pushed at you—so the response is quite different. What Facebook has unleashed will not go away!"

13.

The State of the Media & the Death of Brands

If online social networks in general, and Facebook in particular, are the "Big New Channel," what does that portend for the Big Old Channels—those legacy brands that once dominated our media attention? One reliable indicator is a comprehensive report called The State of the News Media, released annually by the Pew Research Center's Project for Excellence in Journalism. The 2009 edition, which looked back on trends in American journalism in the previous year, began rather ominously by noting, "Some of the numbers are chilling."

The numbers in question involved revenue, valuation, staffing, ratings, circulation and other key metrics. Newspaper advertising revenues, for example, were down 23 percent in just two years. "Some papers are in bankruptcy, and others have lost three-quarters of their value," the Project's staff wrote. "By our calculations, nearly one out of every five journalists working for newspapers in 2001 is now gone, and 2009 may be the worst year yet."

The damage was not limited to newspapers. "In local television, news staffs, already too small to adequately cover their communities, are being cut at unprecedented rates; revenues fell by 7% in an election year—something unheard of—and ratings are now falling or flat across the schedule. In network news, even the rare programs increasing their ratings are seeing revenues fall." Only cable news had flourished, the report explained, "Thanks to an Ahab-like focus on the 2008 election."

Along with the numbers, the report's conclusions were also chilling: "Perhaps least noticed yet most important, the audience migration to the internet is now accelerating. The number of Americans who regularly go online for news jumped 19% in the last two years, according to one survey. . . . It is now all but settled that advertising revenue—the model that financed journalism for the last century—will be inadequate to do so in this one."

With legacy media's advertising and audience both down in every sector but cable and online, the Pew report merely quantified the obvious: the news industry was in "a race against the clock for survival." Would it be possible to use "the declining revenue of the old platforms to finance the transition" to online newsgathering?

Perhaps but the declining ad revenue and audience migration to the Web meant "the news industry has to reinvent itself sooner than it thought." In addition, the collapsing economy caused by the Great Recession had doubled revenue losses. "In trying to reinvent the business, 2008 may have been a lost year, and 2009 threatens to be the same," the report concluded.

"This is the sixth edition of our annual report on the State of the News Media in the United States," the authors intoned. "It is also the bleakest."

Still, there was *some* good news. First, what the report called "the old media" had at least held onto its audience as consumers moved online. Second, "the old norms of traditional journalism" still appeared to have value. Finally, it noted, "consumers are not just retreating to ideological places for news." Yet the trend was clear: "Audiences now consume news in new ways. They hunt and gather what they want when they want it, use search to comb among destinations and share what they find through a growing network of social media."

Despite the modicum of positive signs, however, the state of the news media was still bleak the following year. "What now?" the 2010 Pew report began by asking. What indeed? Resources for traditional news media continued to shrink even as the business slowly emerged from the worst of the recession, which had decimated advertising revenues across the board:

- Newspaper advertising revenue plummeted an additional 26%, bringing the loss over three years to 43%.
- Local television revenue was down 22%—triple the decline of the previous year.

- Radio was also off 22% . . . magazine ad revenue 17% . . . network television 8%—and news programming alone even more. Among the major legacy media sectors, only cable news again escaped declining revenue.

Sand in an Hourglass

In almost any scenario, the economic future for virtually all legacy media looked dismal. Even assuming an economic recovery (one we have yet to see, of course) three elements of the "old media," newspapers, radio and magazines, will take in 41 percent less in ad revenues by 2013 than they did in 2006, according to projections by the market research and investment banking firm Veronis Suhler Stevenson. "For newspapers, which still provide the largest share of reportorial journalism in the United States, the metaphor that comes to mind is sand in an hourglass," the report noted. "The shrinking money left in print, which still provides 90% of the industry's funds, is the amount of time left to invent new revenue models online. The industry must find a new model before that money runs out."

The losses were already enormous. Working with Rick Edmonds of the Poynter Institute, the report's authors estimated that the newspaper industry had lost $1.6 billion in annual reporting and editing capacity in less than a decade, or roughly 30 percent. Staffing at the weekly newsmagazines *Time* and *Newsweek* was down by 47 percent (before the latter was sold at a fire sale price by its *Washington Post* owners, who had already lost nearly 30 million dollars and foresaw only further bleeding.) Meanwhile, network news division resources were down by more than half—in other words, hundreds of millions of dollars—from their peak in the late 1980s, and new rounds of cuts came one after another. (ABC News, to give one example, eliminated tens of thousands of jobs, affecting fully one-quarter of all its employees, in 2010.) Cutbacks, buyouts, layoffs, losses—tens of thousands of jobs, hundreds of millions of dollars—was there no end to the woes afflicting legacy news media?

The short answer was, simply, "No!" Academic analysts like Clay Shirky of New York University suggested the loss of news resources was a predictable gap in a process of creative destruction. "The old stuff gets broken faster than the new stuff is put in its place," Shirky explained. Others like Michael Schudson, a sociologist of journalism at Columbia University, were slightly more optimistic, seeing the promise of "a better array of public informational resources emerging" and a new media ecosystem that would include different styles of journalism, a mix of professional and amateur approaches and, most importantly, different economic models.

To some extent this is already happening. As the report noted, "Twitter and other social media emerged as powerful tools for disseminating information and mobilizing citizens such as evading the censors in Iran and communicating from the earthquake disaster zone in Haiti." The majority of people on the Internet now use some kind of social media, including Twitter, blogging and networking sites, according to another recent PEJ/Pew Internet & American Life survey. Consumers' discovery of legacy-produced news from Twitter, Facebook, and other Internet sources such as Google has forced the "unbundling" of news products, an effect similar to that Apple iTunes had on the music industry when it enabled consumers to purchase single songs instead of acquiring an entire album.

Other notable effects of this unbundling include a loss of importance of the news media's home pages—and with it, the loss of "content the site can vouch for—usually already vetted." The trend was started by Google's "deep linking" to searches for specific stories and later accelerated by social media. As *Bloomberg Businessweek* editor Josh Tyrangiel told me, "Every page has to be a home page now."

With the link economy of both search and social now driving fully two-thirds of all visitors to legacy media sites, usually directly to their article pages, and legacy control of their highly concentrated and centralized distribution networks slipping away, this abrupt change in news-and-information distribution patterns is having a severe impact on news media brands. As Tyrangiel noted, "Facebook is the number one driver of attention, so yes, social networks matter greatly."

Many media leaders still frequently cite brands as the best way to filter through the Internet's supposed cesspool of misinformation and to solve society's growing gap of trust amidst a glut of information, however. For proof, we need only look at network television, says former CBS News President Andrew Heyward.

Heyward, who describes himself as "more of an Eric Schmidt guy on brands as filters," points out that "Share is going down, but CPM [cost-per-thousand, a crucial advertising metric] is going up" at the network level. "They get a premium, because as it becomes harder to reach scale, it becomes more valuable to be big," he notes. "So brands may not be sufficient or even necessary—but they remain damn useful."

Heyward's analysis is shared by Richard Gingras, former CEO of the Salon Media Group. "In a chaotic marketplace such as this, brands are more important than ever," says Gingras. "Chaos increases the value of brands." Gingras, who left Salon in 2011 to become head of news products at Google, foresees a future of "enhanced by both the legacy brand provider and its journalists' individual brands, along with social media's friends and followers, plus algorithms and machines—and now curation is becoming a brand exercise as well." Gingras believes a "mixed brand power will emerge from a hierarchy of trust," and each will be used for different purposes. "Already I go to legacy branded sites daily, since they are a filter, and then I go to Facebook for serendipity beyond the brands," he says. "Brands are a solution, just not *the* solution!"

"Will brands continue to act as filters?" asks leading communications researcher Paul Resnick. "One perspective is that with more and more new brands, we are actually seeing the dilution of brand power. So there could be huge opportunities for micro-brands in the future, along with curators and algorithmic recommender systems."

The future media mix could include algorithmic aggregators and lots of micro-brands, "All of which will stretch the notion of what a brand is," Resnick says. "So Eric Schmidt is right if we expand the notion of brands—but if by brands he means just CBS, Fox and other legacy media . . . no!"

Bloomberg Businessweek editor Tyrangiel adds that it is important to differentiate "between tool and message brands." He believes that media or "message" brands "matter more over time because we have a more emotional identification with them.

"If someone invents a better social network, then Facebook will die," says Tyrangiel. "The same with Google—there is no loyalty whatsoever to tool or technology brands—but there is some to message brands because the importance of their message transcends any medium. Message brands are more consistent, tool brands more volatile. If Google search is somehow eclipsed, people will be gone tomorrow!"

A Set of Friends as a Set of Brands

As entrepreneur and venture capitalist Reid Hoffman, cofounder of the business-oriented social network LinkedIn, points out, "There are lots of brands now. So the *ABC News* may be valuable—but is it ten times or one hundred times more valuable than the next guy on the Net? I don't think so . . . in other words brands will continue to be useful, but the parallel is that new brands are being created every day."

Hoffman says brands are more fluid now. "At the 50,000 feet level, brands look good and they are obviously important," he concedes. "But we need a much broader definition of brands in the future. Maybe we are seeing the death of brands, as we now understand them. We'll need either to make the definition much more broad, or else call them something different, because it's all changing.

"With Web 2.0, anyone who has a presence and identity online can be a publisher," Hoffman points out. "So in a sense, everyone is a brand now. When you're living in a networked world, personal brand management is the future. We need to massively redefine what we mean as brands. Branding is going away as its own thing but will now be integrated into an ecosystem. The death of brands means we will be talking much more to a set of friends as a set of brands."

"The death of corporate brands is a real phenomenon," agrees Internet researcher Ethan Zuckerman, director of the Massachusetts Institute of Technology's Center for Civic Media. "People want to talk to a person, so now brands and institutions are putting a human face to their brand. Traditional gatekeeper brands tell you it's important to know something and have a reasonable chance of it being truthful. As these gatekeepers are lost, this process, which is hard and expensive, gets lost as well. The Net now gives us the opportunity to hammer at the truth, to put it out there, and then correct as necessary.

"Is the Net a cesspool?" asks Zuckerman. "Well, there's a lots of shit out there, sure. But are brands then the solution? Perhaps saying that 'reputation is the solution' is a more sophisticated way to put it. It could be an individual reputation, for example, which we can compare to a brand, since trust will accumulate over time and so on. Eric Schmidt and Google already have a powerful brand, so it is in his self-interest to say what he does. I think brands will be worthless; reputation will matter much more." Even those still firmly in the Eric Schmidt camp, such as Andrew Heyward, say brands "should still be queried and questioned." Heyward notes that there is a "huge disintermediation of brands going on, so we must redefine brands within brands. We have to disaggregate the mainstream media ourselves now." Finally, Heyward adds a warning for corporate leaders outside the media world. "What's happening to the new media is definitely a harbinger for all other industries," he says. "There is no doubt about it, and trust is central to it all."

So whether you are in charge of a tool or a message brand, a media corporation or any other, be advised: the digital information revolution is leading not only to the obsolescence of your home page, but inevitably to the diminution, if not actual death, of your once-valuable brand as well—something that will soon become apparent even to outliers such as Google's Eric Schmidt.

At the moment, however, it's the legacy news media that are most immediately imperiled by the changing nature of brands. As the 2010 Pew study detailed, consumers no longer seek out them for a full news agenda. Instead, they are "hunting the news by topic and by event and grazing across multiple outlets." At the same time, citizen journalism is expanding and brimming with innovation, as are new, specialized news websites such as the Global Post, ProPublica, and Kaiser Health News, to name just a few.

It's gratifying that ProPublica and other new non-profit media outlets such as the *Voice of San Diego* and the *Texas Tribune* are helping to preserve what Alex Jones, in his book *Losing the News*, called "the iron core" of journalism, and filling part of the gaping hole resulting from the endless rounds of revenue challenges facing the commercial media world. But it's also apparent, as the Pew report concludes, "the scale of these new efforts still amounts to a small fraction of what has been lost." As one newspaper veteran told the Project's staff, "All the metrics for legacy news media are headed in the same direction: less revenue, far fewer reporters and editors, less profit to reinvest and less community clout."

The old mass audience newspaper, for example, is fast becoming a narrow niche newspaper with an audience of old and wealthy readers, as the web and its links now attract the masses. Along with advertising, publishers have seen circulation slide: in 1998, 45 percent of U.S. households received a daily paper; in many cities today, print penetration now approaches just 30 percent. At the same time, digital businesses contribute no more than 15 percent of publishers' total revenues.

By the time the Project for Excellence in Journalism released its 2011 report, the state of the legacy media had improved . . . somewhat. Revenue began to recover slowly for most sectors, although newspapers continued to suffer declines—"an unmistakable sign that the structural economic problems facing newspapers are more severe than those of other media," according to the report's authors, Tom Rosenstiel and Amy Mitchell.

Fundamental challenges still remained, however, most notably that the legacy media no longer control their own future and are forced to rely instead on independent networks to sell their ads and "on aggregators (such as Google) and social networks (such as Facebook) to bring them a substantial portion of their audience." In this new media world "where consumers decide what news they want to get and how they want to get it," said Rosenstiel and Mitchell, new intermediaries stood between the old media and their audience, which continued its migration to the web.

In 2010, as they noted, "every news platform saw audiences either stall or decline—except for the web." The Internet now trails only television as a news destination, and that gap is closing. Even cable news viewer ratings shrank for the first time in years. "Financially the tipping point also has come," the report said. "When the final tally is in, online ad revenue in 2010 is projected to surpass print newspaper ad revenue for the first time. The problem for news is that by far the largest share of that online ad revenue goes to non-news sources. . . ."

With their content aggregated on the web and their home pages hurt by direct links from search and, increasingly, social media, legacy operations are increasingly disconnected from their readers and viewers, who in turn increasingly disconnect from their previous reliance on the dying legacy brands. Meanwhile, web users have made Facebook and Twitter their online hangouts, spending hours there monthly but no more than 10 to 15 minutes on news websites. Social networking now takes up twice as much time as any other online activity. According to the latest statistics from Nielsen, sites like Facebook and Twitter now account for 22.7 percent of time spent on the web; the next closest activity is online games at 10.2 percent. In this new media world, old brands are rapidly waning in their reach, their meaning and their importance in assembling audiences.

As Jeff Jarvis wrote on his buzzmachine blog, "Back in the day, we discovered media—news, information, or service—through brands: We went and bought the newspaper or magazine or turned on a channel on its schedule. That behavior and expectation was brought to the Internet: Brands built sites and expected us to come to them."

Now, as Jarvis notes, there are "other spheres of discovery." Once brands decided what we should want and succeeded largely because they controlled distribution of content. "One-way, one-size-fits-all, fleeting—these are the characteristics of branded media," Jarvis says. Now brands are being disrupted by search, by social, by links and by new tools and technology, which "disaggregate and disintermediate them, challenge their authority, compete on much lower cost bases (thanks to automation and collaboration), and provide better targeting and relevance (thanks to new means of gathering and analyzing data)."

Obviously brands still play important roles in the current media ecology, with its confusing plethora of content and persistent questions of credibility. "People are always looking for trust shortcuts," says media researcher Kelly Garrett of Ohio State University. "It's either brands, some sort of credential, or some sort of social network—but they are making up their own ways of trust assessment." Stanford's BJ Fogg agrees, "Brands can be shortcuts," but adds that they are rapidly losing prominence.

"The mainstream media *had* a sort of trusted brand—but they've given up a lot of trust of late," Fogg says. "The issue around brands is that different friends trust different brands. The challenge now is that there are no destination sites, so that undercuts the value of news brands. And lost trust equals a lost brand." Fogg believes that the legacy media "deserves what has happened to them—and once you lose credibility, it's very, very hard to regain. It's hard to change people's habits—especially the young—once that trust and that brand is damaged."

The bottom line: legacy media brands, while still quite powerful, are busy dying; they can no longer expect consumers/viewers/readers/users to come to them but remain incapable of anticipating audience needs by listening to their signals. As we no longer trust brands, we will instead discover more and more content through *people* and *networks* we trust.

14.

Politics 3.0

On December 17, 2010, a young Tunisian street vendor named Mohamed Bouazizi set himself afire. Within months, much of the Arab world was ablaze as well.

Bouazizi's fatal self-immolation sparked what later became known as the Arab Spring and led to the demise of dictators not only in his country, but in neighboring Egypt and Libya as well. It reverberated among other repressive rulers and regimes in Syria, Bahrain, Yemen, and beyond. Ultimately it resonated far from Bouazizi's rural hometown of Sidi Bouzid—including thousands of miles away in New York City, where Occupy Wall Street movement began nine months to the day after his desperate act in opposition to voicelessness and powerlessness. The Occupy movement gave rise in turn to more than a thousand similar protests around the world and the creation of a global movement. From London, Madrid and Rome to Athens, Tel Aviv and Tokyo, millions were suddenly on the march, demanding, as had Bouazizi, more respect, hope, dignity and democracy.

They called themselves "the 99%" in opposition to the

ruling 1%—the dictators like Ben Ali, Mubarak, and Gaddafi, but also those they dubbed "banksters," the investment bankers, and financial manipulators who had gamed the economic system to their own benefit and then, supposedly too big to fail, forced the rest of society to bail them out. On the heels of the Great Recession, the American Autumn, like the Arab Spring that preceded it, became as much an economic as a democratic revolt. As one writer, Rebecca Solnit, later asked in an open letter to the dead Bouazizi, "What were all those dictatorships and autocracies for, if not to squeeze as much profit as possible out of subjugated populations—profit for rulers, profit for multinational corporations, profit for that 1%?"

Mohamed Bouazizi was neither the first Tunisian to kill himself in protest of the economic and political conditions in his country nor the last. But something was different about his story, or at least the way it had been told. Bouazizi's personal rebellion led first to a successful societal revolution against a man who had ruled Tunisia for decades with an iron fist, and later to others that dislodged dictators in Egypt and Libya. Why? Did it have anything to do, as some have suggested, with the rise of social media?

Not long after the fall of the Tunisian dictator, many observers, including U.S. President Barack Obama, proclaimed that social media had in fact played a key role in events there, as well as in Egypt, where strongman Hosni Mubarak's regime was also toppled. Writing for the Reuters news agency, for example, Philip N. Howard, author of *The*

Digital Origins of Dictatorship and Democracy: Information Technology and Political Islam, noted that Bouazizi's death had "activated a transnational network of citizens exhausted by authoritarian rule. . . . It was social media that spread both the discontent and inspiring stories of success from Tunisia across North Africa and into the Middle East." Among the lessons for the West, Howard concluded, were the facts that "a larger network of citizens now has political clout, largely because of social media," and that "democratization has become more about social networks than political change driven by elites."

Months later, after analyzing more than 3 million tweets, gigabytes of YouTube content and thousands of blog posts, he and other scholars at the Project on Information Technology and Political Islam published a study claiming "social media played a central role in shaping political debates in the Arab Spring," and noting that, "conversations about revolution often preceded major events on the ground, and social media carried inspiring stories of protest across international borders."

Howard, an associate professor of communication at the University of Washington, said the evidence "suggests that social media carried a cascade of messages about freedom and democracy across North Africa and the Middle East, and helped raise expectations for the success of political uprising. People who shared interest in democracy built extensive social networks and organized political action. Social media became a critical part of the toolkit for greater freedom."

During the week before Mubarak's resignation, for example, the rate of tweets about political change in Egypt increased ten-fold and videos featuring protest and political commentary went viral, with the top two- dozen receiving nearly five and a half million views. The amount of content produced in Facebook and political blogs by opposition groups also increased dramatically. Ironically, government efforts to crack down on social media may only have incited more activism, especially in Egypt. People who were isolated by efforts to shut down the Internet, largely middle-class Egyptians, may have gone to the streets when they could no longer follow the unrest through social media.

"Recent events show us that the public sense of shared grievance and potential for change can develop rapidly," Howard concluded. "These dictators for a long time had many political enemies, but they were fragmented. So opponents used social media to identify goals, build solidarity and organize demonstrations."

Other researchers and scholars, however, are not so sure of the actual role social media played in facilitating the protests. Writing in mid-September, 2011 on nextgov.com, a website devoted to "technology and the business of government," Joseph Marks reported that although experts were in agreement that, "Something extraordinary happened at the nexus of social media and political action during the Arab spring uprisings in the Middle East and North Africa . . . just what happened is less clear." While Twitter and other social media had become a megaphone that disseminated information about the uprisings to the outside world, Marks said, "a comprehensive study of Tweets about the Egyptian and Libyan uprisings" between January and March found that more than 75 percent of people who clicked on embedded Twitter links related to the uprisings were from *outside the Arab world.*

As one researcher, GWU associate professor John Sides, noted, "This obviously suggests that new media presents a tremendous opportunity to inform an international audience, but it also raises the question: 'Will they be there tomorrow?'" Sides said public attention spans in the Western world are limited and cited Iran's 2009 Green Revolution as an example. Although the Iranian events attracted a surge of international activity on Twitter, attention dwindled shortly after the death of pop icon Michael Jackson (see Chapter 7).

Alec Ross, senior adviser for innovation at the U.S. State Department, supported the idea that social media had played a determinant role in the Arab Spring. Ross said the use of social media during the uprisings signaled the beginning of a "massive transfer of power from nation states and large institutions to individual and small institutions." Other panelists warned, however, that "data on the role of social media during the Arab spring is so disparate and confusing it is nearly impossible to draw meaningful conclusions from it."

Cyber-Realists vs. Cyber-Utopians

The ongoing controversy over whether and to what extent social media helped create the democratic surge of the Arab Spring brought to the fore earlier disagreements between "cyber-utopians" and more skeptical "cyber-realists" such as Malcolm Gladwell and Evgeny Morozov, author of *The Net Delusion: The Dark Side of Internet Freedom*. Gladwell's *New Yorker* article, headlined "Small Change: Why the Revolution Will Not Be Tweeted," created a storm of reaction—most of it negative—when it was published two months before Mohamed Bouazizi's death.

"The world, we are told, is in the midst of a revolution," Gladwell had written. "The new tools of social media have reinvented social activism. With Facebook and Twitter and the like, the traditional relationship between political authority and popular will has been upended, making it easier for the powerless to collaborate, coordinate, and give voice to their concerns." Like Morozov, Gladwell was convinced that those he derided as "digital evangelists" had, at the very least, vastly overstated the impact of social media on the new wave of political activism.

As evidence he cited reaction to the protests in both Iran and Moldova in 2009. When thousands of demonstrators took to the streets in Moldova against their country's government, Gladwell noted, "The action was dubbed the Twitter Revolution, because of the means by which the demonstrators had been brought together." And when protests later erupted in Tehran, the U.S. State Department asked

Twitter executives to suspend previously scheduled maintenance of the service so it could still be used as an organizing tool during the demonstrations.

Gladwell remembered derisively that former U.S. national-security adviser Mark Pfeifle had called for Twitter to be nominated for the Nobel Peace Prize and had said, "Without Twitter the people of Iran would not have felt empowered and confident to stand up for freedom and democracy." He also recalled former U.S. State Department official James K. Glassman telling a crowd of activists that sites like Facebook "give the U.S. a significant competitive advantage over terrorists. Some time ago, I said that Al Qaeda was 'eating our lunch on the Internet.' That is no longer the case. Al Qaeda is stuck in Web 1.0. The Internet is now about interactivity and conversation."

These were "strong, and puzzling, claims," Gladwell said. After all, "Why does it matter who is eating whose lunch on the Internet?" Like Morozov, Gladwell believed "Moldova's so-called Twitter Revolution" was impossible since very few Twitter accounts exist there. As for Iran, the people "tweeting about the demonstrations were almost all in the West."

Writing in *Foreign Policy*, Radio Free Europe correspondent Golnaz Esfandiari supported Gladwell's conclusion. "Simply put: there was no Twitter Revolution inside Iran," Esfandiari stated forthrightly. Twitter's impact inside Iran was nil, she believed, as did the manager of one of the Internet's most popular Farsi-language websites, Mehdi

Yahyanejad, whom Esfandiari quoted as saying, "Here [in the United States,] there is lots of buzz. But once you look, you see most of it is Americans tweeting among themselves." Those who disagreed, Esfandiari continued, were lazy and uninformed. "Western journalists who couldn't reach—or didn't bother reaching?—people on the ground in Iran simply scrolled through the English-language tweets post with tag #iranelection," she wrote. "Through it all, no one seemed to wonder why people trying to coordinate protests in Iran would be writing in any language other than Farsi."

Grandiose claims for new media forms were only to be expected, Gladwell concluded. "Innovators tend to be solipsists. They often want to cram every stray fact and experience into their new model." But there was something else at work, as well: "in the outsized enthusiasm for social media . . . we seem to have forgotten what activism is."

Two months after Bouazizi's death, amid a new spate of claims that the protests in Tunisia and Egypt were also Twitter or alternately, Facebook-inspired, Evgeny Morozov decried as "cyber-utopians" those who believe "the Arab spring has been driven by social networks." In a post for the UK *Guardian*, Morozov argued that they "ignore the real-world activism underpinning them."

Morozov began his argument by restating theirs, *reductio ad absurdum*, "Tweets were sent. Dictators were toppled. Internet = democracy. QED." He found it sad but entertaining, he said, to watch as "adherents of the view that digital tools of social networking such as Facebook and Twitter can summon up social revolutions out of the ether trip over one another." He also complained about what he called "the on-going persecution of Malcolm Gladwell."

Like Gladwell, Morozov is convinced, "The current fascination with technology-driven accounts of political change in the Middle East is likely to subside, for a number of reasons." First, he said, accounts of the revolutions that emphasize the liberating role of social media tools function mostly to "make Americans feel proud of their own contribution to events in the Middle East. After all, the argument goes, such a spontaneous uprising wouldn't have succeeded before Facebook was around—so Silicon Valley deserves a lion's share of the credit."

Second, he suggested that social media "by the very virtue of being 'social'—lends itself to glib, pundit-style over-estimations of its own importance." He then added, with no small degree of snark, "Perhaps the outsize revolutionary claims for social media now circulating throughout the West are only a manifestation of western guilt for wasting so much time on social media: after all, if it helps to spread democracy in the Middle East, it can't be all that bad to while away the hours 'poking' your friends and playing FarmVille."

Morozov, Gladwell and their allies cloak their argument in academic terms often used to describe different types of social capital within networks. They believe what they call "high-risk, real world activism," such as that demonstrated during the Arab Spring, to be a "strong-tie" phenomenon.

"The kind of activism associated with social media isn't

like this at all," Gladwell wrote. "The platforms of social media are built around weak ties."

Social networks are "simply a form of organizing which favors the weak-tie connections that give us access to information over the strong-tie connections that help us persevere in the face of danger," he said, making it "easier for activists to express themselves, and harder for that expression to have any impact.

"There is strength in weak ties.... Our acquaintances—not our friends—are our greatest source of new ideas and information," Gladwell conceded. But those same weak ties, "seldom lead to high-risk activism," he concluded. "The evangelists of social media don't understand this distinction; they seem to believe that a Facebook friend is the same as a real friend."

Social networks, these naysayers claim, are ill-suited to real-world activism and high-risk strategies such as those employed during Arab Spring—boycotts and sit-ins and nonviolent confrontations—because they are messy, non-hierarchical, and cannot provide the necessary discipline and strategy. When taking on a powerful and organized establishment, Gladwell declared, "You have to be a hierarchy."

The explanation for why Mohamed Bouazizi's death uncorked such a fury of change, however, is more nuanced than either of the dueling cyber-camps is willing to admit, as a closer examination of the protests in Tunisia and later in Tahrir Square in Cairo as well as the various Occupy demonstrations that followed in other parts of the world, seems to suggest. In the North Africa/Middle East region, a pan-Arab collaboration of young activists skilled in the use of technology did in fact given birth to a new movement dedicated to spreading democracy. They were strategic and disciplined, even as they shied from hierarchy. They relied not only on tactics of nonviolent resistance but also those of marketing borrowed from Silicon Valley. Tunisians and Egyptians did share expertise and experiences with similar youth movements in Libya, Algeria, Morocco and Iran. "Tunis is the force that pushed Egypt, but what Egypt did will be the force that will push the world," Walid Rachid, one of the members of the April 6 Youth Movement, which helped organize the protests that set off the Arab Spring, explained to the *New York Times*.

That being said, it is also true that both the Tunisian and Egyptian revolts were literally decades in the making. Speaking in June 2011 at the eight annual Personal Democracy Forum, a procession of young Arab activists who had all been intimately involved in the spring revolts, explained the process of how they and millions of supporters were "weaving a network for change" in their countries, and what role the emerging media played in making that happen. From Riadh Guerfali to Dr. Rasha Abdulla to Mona Eltahawy to Alaa Abdel Fattah, they noted the Arab Spring actions were emphatically not "Twitter" or "Facebook" revolutions that had coalesced online, but were instead the outcome of decades of networked resistance *offline*.

At the same time, they said, the revolts were clearly

facilitated, and to some extent accelerated, by the decentralized organizing power of the new social media. The results of this offline/online action mashup were surprisingly successful revolutions that overthrew long-entrenched political forces. As Alaa Abdel Fattah pointed out in his remarks at the forum, the roots of the revolution in Egypt went back as far as 1972 and efforts made by his parents' generation. Ultimately, he explained, they had been stymied by a clever power structure that painstakingly divided and thus conquered the protesters, marginalizing some and buying others off with favor and access. Decades later, Fattah pointed out, the emerging social media suddenly made it possible "to make noises louder online, to build local movements with one narrative and then build them online to a mass movement." As another speaker at the forum, Omoyele Sowore, explained, "The Internet has helped revolution; but the Internet is not revolution."

The Human Router

Evidence that the Internet had indeed helped the revolutions in Tunisia, Egypt, and, to some extent, Libya can be found by examining the aggregating and curating activities of Andy Carvin on Twitter. A social-media strategist at NPR in Washington, Carvin became what Paul Farhi described in a *Washington Post* profile as "a one-man Twitter news bureau, chronicling fast-moving developments throughout the Middle East. By grabbing bits and pieces from Facebook, YouTube and the wider Internet and mixing them with a stunning array of eyewitness sources, Carvin has constructed a vivid and constantly evolving mosaic of the region's convulsions."

Carvin started tweeting about the Middle East during the beginning of the rebellion in Tunisia in December 2010. As he later explained in an interview with Melissa Bell of the *Post*, he knew bloggers there from earlier trips, "though they generally focused on tech blogging rather than anything political." As he began to see people on Twitter using a hashtag of #sidibouzid—Mohamed Bouazizi's hometown—to document and encourage protests, he wondered, "if there was any chance they would be able to go all the way." Ultimately, they did—and Carvin and tens of thousands of his followers on Twitter went with them, as he documented the progress of the rebellion using the curation tool Storify, which became his "own record of the whole revolution, told through social media created by Tunisians and others."

On the day that Tunisian president Ben Ali fled, Carvin saw a tweet from a Tunisian blogger challenging others in the region to act, "Ok Arabs you've seen how it's done in Tunisia; Tag you're it!" Soon he also saw Egyptian bloggers using a new hashtag—#jan25—representing the day they planned to begin their own protests. He began putting together lists of all the Egyptians he knew on Twitter and filtering their tweets as well. The result, as the *Washington Post* noted, was "a dizzying, nonstop ride across the geopolitical landscape, 140 characters at a time. . . . Seven days a week, often up to 16 hours a day."

For weeks, Carvin could be found "tweeting links to fresh video from Libyan rebels, photos of street protests in Bahrain or the highlights of a NATO news conference," Farhi noted. "His followers, in turn, point him to more material—on-the-ground accounts of the government crackdown in Yemen, breaking reports from Tahrir Square, the latest from Jordan or Syria."

What Carvin did was so groundbreaking the words to describe it didn't exist yet—was he a "tweet curator? a social-media news aggregator? an interactive digital journalist? or a human router?" reporters variously wondered—but by "storifying" Twitter to report, fact check, and encourage his followers to help translate, find more information and confirm rumors, he was able to provide what many found to be "the most consistent, thorough, and wide ranging aggregation of the events in the Middle East."

In the process, Carvin demonstrated the extent to which social media was helping to make real-world changes in political institutions around the world. "It's been incredible watching people find their public voices in countries where they could be thrown in jail or worse for doing that," he told the *Post*'s Melissa Bell. "Free speech and free press generally hasn't been respected in these countries before, so it's amazing to watch people discover their voice and use it. I've even started to notice that some of the people I've been following in Egypt have taken down their anonymous Twitter avatars and replaced them with actual pictures of themselves. That in itself says a lot about how much their world has changed. They can show their faces and speak up for the first time in their lives."

Further evidence of social media's power to stimulate real-world, high-risk activism could be found many miles away and months later, as a small group of protesters gathered in a park in New York City in September 2011 and sparked an online conversation that began spreading across the country on social media platforms. Inspired by the message of the New York group, which soon became known the world over as Occupy Wall Street, hundreds of Facebook pages and Twitter accounts were created in just a week, as similar protests spread to more than one thousand other locales around the globe.

"We are not coordinating anything," Justin Wedes, a former high school science teacher from Brooklyn who helped manage one of the movement's main Twitter accounts, told Jennifer Preston of the *New York Times*. "It is all grass roots. We are just trying to use it to disseminate information, tell stories, ask for donations and to give people a voice."

Mark Ghuneim, founder and chief executive officer of the social media analytics firm Trendrr, told Preston that the conversation on Twitter was producing an average of 10,000 to 15,000 Occupy movement posts an hour, with most people sharing links from news sites, Tumblr, YouTube and elsewhere. "This is more of a growing conversation than something massive as we have seen from hurricanes and with people passing away," Ghuneim said. "The conversation for this has a strong and steady heartbeat that is spreading. We're seeing the national dialogue morph into pockets of local and topic-based conversation."

As Preston noted, "In Egypt, the We Are All Khaled Said Facebook page was started 10 months before the uprising last January to protest police brutality. The page had more than 400,000 members before it was used to help propel protesters into Tahrir Square. Occupy Wall Street's Facebook page began a few weeks ago and has 138,000 members."

But that page on Facebook was only one of many. Other localized Occupy pages soon amplified the issues being discussed in New York. Thousands of YouTube videos tagged "occupywallstreet" were uploaded as well. Soon protesters all over the country were carrying signs with the movement's message, "We are the 99 percent." By November 2011 the movement had become a startling new force in American political life and sparked similar protests all over the world.

The trend came full circle when Egyptian social media activists Esraa Abdel Fatah, Ahmed Maher, and Bassem Fathy—a founder of April 6 Youth, which Salon's Jefferson Morley noted had "used Facebook, Twitter and YouTube to detonate a social explosion that swept away Mubarak's government last January"—appeared at an OccupyDC protest in Washington's Freedom Plaza.

"As the crowd plied their guests for advice about how the U.S.-based occupation movement should proceed," Morley reported, "the Egyptians responded by voicing the unorthodox tenets of a global movement without leaders or unified set of demands." Fatah counseled, "People will want to know who your leaders are. . . . Your demands must be your leaders." Maher added, "My advice is not to accept any advice."

That sounds like good advice. . . . After all, in our media-saturated age, hope and hype often go hand-in-hand. One reason the emerging media tools attracted so much attention—from the protests in Cairo's Tahrir Square to the Occupy Wall Street demonstrations in New York's Zuccotti Park and beyond—is that many legacy media reporters still struggle to comprehend the mobilization and change these new tools appear to engendered.

But let's also be clear: social media, like all media before them, can be used by institutions as well as individuals, and either for good or for ill, in order to transmit lies or truth and to promote or constrain liberty. The same tools employed by the relatively powerless 99% are being used on behalf of the powerful 1% in repeated attempts to hold back the tide of change. In Iran, for example, government forces tried to subvert the rebellion there in 2009 by employing social networks to plant false information and ferret out opposition. In Egypt, the Mubarak administration even went so far as to shut down the entire Internet in a desperate attempt to maintain control. In New York, the police department uploaded its own videos of Occupy Wall Street protests onto YouTube, including those from a demonstration that led to 700 arrests, in an attempt to counter the use of social media by protesters there.

The CIA's Ninja Librarians

As The Associated Press revealed in November 2011, several hundred analysts from the CIA's Open Source Center, they call themselves "ninja librarians," now look at as many as 5 million tweets a day, "mining the mass of information people publish about themselves overseas, tracking everything from common public opinion to revolutions."

The Center started focusing on social media and poring "over tweets, Facebook, newspapers, TV news channels, local radio stations, Internet chat rooms—anything overseas that anyone can access, and contribute to, openly—after watching the Twitter-sphere rock the Iranian regime during the Green Revolution of 2009," the AP noted, "When thousands protested the results of the elections that kept Iranian President Mahmoud Ahmadinejad in power." The CIA center "gives the White House a daily snapshot of the world built from tweets, newspaper articles and Facebook updates," according to its director, Doug Naquin, who said his team had "predicted that social media in places like Egypt could be a game-changer and a threat to the regime."

So both the President of the United States and the CIA are convinced that new platforms like Twitter, Facebook, YouTube and many others can play a catalytic and highly disruptive role in creating political change, particularly in repressive societies where access to other media has long been severely circumscribed. Perhaps using such derogatory and divisive language as "cyber-utopians" while analyzing the ways social media informs and mobilizes people is unhelpful.

Social media obviously is neither a panacea for the world's problems nor a substitute for "real-world activism." It isn't the only means of communication being used in the struggle either, although judging from many legacy media reports, one would be tempted to think so. In addition to relying on social media to tell their story, protesters in New York, for example, published a highly professional and useful broadsheet newspaper cleverly dubbed *The Occupied Wall Street Journal* and created well-designed websites that served as clearinghouses for information and aggregated news about occupations from all over the world.

There should be little doubt, however, that social media are at the very least important tools that favor those that did not have prior access to the means of media production and distribution, or that they help to change the balance of power somewhat. At the same time, it's easy to exaggerate their effect. Ultimately, it's people who make the difference—and not technology. Social media are not important in and of themselves but because of the processes they lead to and their innate tendency to increase "the capacity of citizens to freely aggregate, share, and create common value together," as Michel Bauwens put it in a post on the PolicyMic.com website. "It is this ability to share, to learn together, to construct social artifacts together, outside the bonds of kinship, salary, and commodity," Bauwens noted, "which undermines both the conventional and neoliberal narratives."

This share and share alike ethos is precisely what emerged in Spain with the creation of the first democratic

"General Assembly" method of decision-making and at the Occupy movement's protests with their invention of the unamplified "human microphone" process of discussion, as people began to rediscover half-forgotten human values they seemed to long for.

As Rebecca Solnit concluded in October 2011in her moving open letter to Mohamed Bouazizi, "The process is also the goal: direct democracy. No one can hand that down to you. You live direct democracy in that moment when you find yourself participating in civil society as a citizen with an equal voice. Put another way, the Occupiers are not demanding that something be given to them but formulating something new. That it involves no technology, not even bullhorns, is itself remarkable in this wired era. It's just passionate people together—and then Facebook, YouTube, Twitter, text messages, emails, and online sites like this one spread the word. . . ."

Remember, "The Internet has helped revolution; but the Internet is not revolution."

Going Forward: The Feed Is Your Friend

As this book goes to press in early 2012, the trends detailed in it, particularly those directly affecting media, commerce and politics, are only accelerating. Social media continue their inexorable rise, and play an ever-growing role in our lives. They now account for nearly one quarter of all the time Americans spend online—the leading category by far, compared with less than ten percent for online games or email respectively. At the same time legacy media are still in decline, as measured by both audience reach and advertising revenue, and the sweeping technological shifts and the unprecedented rupture in the long partnership of news and advertising brands that have been roiling them for years continue unabated, with the disruption still most notable in print media such as newspapers and magazines.

Meanwhile the media brand Americans spend the most time with by far is Facebook, the Internet's most ambitious, technologically sophisticated, and fastest-moving company. Its eight hundred million users—one in every nine people on Earth—spend an average of fifteen and a half hours a month on the site. The monthly usage of all news media sites, by comparison, averages only five to twenty minutes per month, while the total digital advertising revenues of all American daily newspapers is just $3 billion, compared to Facebook's individual share of $2 billion.

For all its success, however, Facebook also continues to infuriate large numbers of its users, as an unending series of poorly communicated changes to the site fuels further controversy. (The latest overhaul was announced at Facebook's fourth so-called "F8" event—its key gathering for developers, press, and the public—where Bret Taylor, Facebook's chief technology officer, proclaimed it "the biggest change we've made to our platform since we launched it at the first F8.") At the same time questions about the privacy implications of Facebook's vast presence on the Web—executives were forced to defend their practice of tracking every page users visit even after they have logged out of Facebook— also dog the company.

Just five years after its public launch, Twitter continues to grow in both size and importance too, delivering 350 billion tweets and signing up more than 600,000 new users every day. For its part, YouTube is seeking new and better ways to serve its 490 million unique monthly users, who now spend a staggering total of 2.9 billion hours per month on the site. Still struggling not to be completely overwhelmed with unedited content, YouTube executives have recently partnered with a variety of outside entities to create and curate videos for one hundred new channels that will feature regularly scheduled programming on such broad themes as fashion, news, and sports.

Legacy brands of all sorts continue to fragment and falter, particularly those of the news media, as the advertisers that once supported them increasingly decamp for social media in general and Facebook in particular, where the same recent changes that dismayed users are considered brand-friendly. And Google—once the leading global brand—continues to lose its buzz. A November 2011 front page story in the *New York Times* headlined, "Google's Chief Works to Trim a Bloated Ship," detailed how the company's " midlife crisis . . . threatens to knock it off its perch as the coolest company in Silicon Valley." Reporter Claire Cain Miller dismissed Google as "an aging giant . . . being pushed around by government regulators and competitors like Facebook." And as AP Technology Writer Michael Liedtke reported, even Google's former CEO Eric Schmidt now admits, "I screwed up," in not pressing the company to focus more on mounting a challenge to Facebook. "I think the industry as a whole would benefit from an alternative," Schmidt added.

On the political front, elections are breaking out all over as we head into another year of historic change. In the Middle East, the promise of the Arab Spring is being put to the test of ballots in lieu of bullets, as Facebook revolutionaries transform themselves into more traditional candidates for public office. Here in the United States, politicians from both major parties are embracing social media as never before. While Barack Obama tries to rebuild the grass-roots movement that propelled him to the White House in 2008 by employing everything from YouTube to Twitter, where he has over 10 million followers, to Facebook, where he boasts 19.3 million friends—chief among them Mark Zuckerberg, whom the president has assiduously courted—Republicans say they are better prepared than ever to compete online in the 2012 contest.

"The notion that the Internet was owned by liberals, owned by the left in the wake of the Obama victory, has been proven false," says Republican political online strategist Patrick Ruffini, who points out that many Republicans in the House and Senate now use social media tools more than Democrats. "It is not necessarily that Democrats or young people or liberals have become less active," notes Aaron Smith, the author of a study on the subject by the Pew Research Center for the Internet and Society. "It is more that older adults, conservative voters and Tea Party activists have come to join the party." And Andrew Rasiej, co-founder of the influential *TechPresident.com* blog, says, "This will be the first election in modern history that both parties are understanding the potential of the technology to change the results of the election. Both Republicans and Democrats are ready to use online platforms and are no longer skeptical of its potential."

Local media strategies will be key to both sides in America's 2012 national election. Social media will be the difference maker, since strategists have figured out how to harness the Internet for hyperlocal purposes. "With old media tools, local press, radio and TV, it was difficult for a candidate to wage a nationwide, local strategy," says one media analyst, Brooklyn Law School's Jonathan Askin. "The Internet finally makes local campaigning, with national themes and local messaging, effective for presidential politics." It may not matter, however, if we are entering a "period in politics that's sort of fact free," as former President Bill Clinton recently warned. . . .

Meanwhile, the Occupy movement in the United States and its international counterparts all over the world continue to agitate, demonstrate, and aggregate in number and influence. They are not content simply with bypassing legacy media corporations to spread their message but have also begun developing new media tools for future use. Examples include the "I'm Getting Arrested" app, which alerts legal support and family via text messaging when a protester is getting arrested, Occupy The Hub, a website for aggregating video feeds, tweets and live chats to provide one-stop coverage of what's going on in the movement, and OccupySMS, a program that allows one person to send a text message quickly and easily to a huge mass of people. "With this sort of innovation, the Occupiers won't need major media companies to get the word out," as Benish A. Shah, vicepresident of Strategic Digital Media at the Global Executive Board, told the Huffington Post. "The people following this are going to go online and find the information and find it from other sources."

Finally, the single most important new trend within the

digital information revolution—the exponentially increasing amount of unvetted and unverified information now washing over us all—continues to flashflood forward at a frightening pace. What's worse, it's harder than ever to tell which waves in the torrent might carry relevant and trustworthy news and information.

A Pew report released in September 2011 showed that only one-quarter of those surveyed think news organizations get the facts right—a new low since the question was first asked in 1985. Two-thirds say stories are often inaccurate—a new high—and nearly three-quarters believe that journalists try to cover up their mistakes rather than admit them. For the first time in any such survey, as many people say that news organizations hurt democracy (42 percent) as protect democracy (42 percent). The situation is so dire that the MIT Center for Civic Media has begun to pursue the development of a "nutritional label for news," which would "semi-automate evaluating the quality of an article" and could be visualized "as easily as an FDA 'recommended daily amount' nutritional label for food."

Everyone Wants To Be a News Filter

All the while, the debate over the "Daily Me" vs. the "Daily We" still rages. As academic media researchers argue over how much they really know, our privacy continues to vanish, unwanted personalization thrives and disputes continue over how best to filter and sort through it all. What are the best means and mechanisms of dealing with the twin crises of too much information and credibility-and-trust: the old, time-tested brands? The new recommender systems and algorithms? Curators and influencers? Friends and followers? Or an adroit mix of all of the above?

"Everyone wants to be a news filter now," Mathew Ingram wrote in a post on GigaOm. "As the avalanche of information coming through social networks and real-time tools like Twitter continues to grow, the need for filters to make sense of that tsunami of data also increases, and it seems as though everyone has a different way of trying to solve that problem." As Ingram noted, however, "Relevance is a tricky problem to solve." Many new apps and approaches suffer from similar problems: "Either they are filled with the same content I've have already seen in other places, or the links simply aren't relevant.

"It's good that plenty of services are trying to solve the news-filtering problem, and different users may choose different solutions," Ingram concluded. "So far, no one seems to have come up with the one-size-fits-all solution to this modern dilemma."

Leading communications researchers remain optimistic, however, particularly about the still-developing roles of both social media and algorithms and learning machines. The University of Michigan's Paul Resnick is among those who remind us that we are still in an experimentation phase. "One key is to develop algorithms that give people what they really want and not the current naive version of it, i.e. what they 'like,'" says Resnick. "Instead of just popularity, there is an opportunity to give people something that will take into account that we have preferences that are sets of items that will engage us the most."

The machines must move away from measurements of mere popularity to become more multidimensional, says Resnick "to consider sets of items over individual items, and to offer us crossover, or 'strange bedfellows' items. We need more sophisticated models of why people want what they want—and we also need consciousness raising among developers."

MIT's Ethan Zuckerman agrees. "With machine learning, the problem now is it produces echo chambers, which are a comfortable filter but may lead to personalization and homophily," he warns. "But in the future, we'll get better systems. In addition, everyone is looking for curators now, and each has own pronounced point of view, so we'll have to learn to 'read the net.'"

Like Resnick, Zuckerman is ultimately hopeful that progress can and will be made in the ongoing effort to separate signal from noise in the crowded and chaotic news-and-information environment. "In diverse enough worlds, we will be able to triangulate our way to the truth," he says. "The real question is how to rebuild institutions—gate keepers if you will—who can tell us the difference between what is credible, relevant and trustworthy and what is not."

Maralee and Me

As mentioned, I began the research for this book in September 2008 while a Fellow at Harvard University's Shorenstein Center on the Press, Politics and Public Policy. Among the other journalists then at the Kennedy School was a woman named Maralee Schwartz. Schwartz was a self-described "lifer" at the *Washington Post*, having spent her entire professional career there. Beginning as a researcher in 1979, she quickly found herself concentrating on politics, working with and learning from such top reporters as David Broder, David Balz and others. By 1992 Schwartz had been named Congressional Editor and four years later she became National Political Editor, responsible for all national political coverage.

As the *Post*'s political editor, Schwartz led an award-winning team of reporters in covering three presidential elections, the last term of the Clinton White House, and the first term of the Bush White House. She held the job for nearly nine years, a record tenure, during which time she directed coverage of the Clinton impeachment, the government shutdown, the disputed 2000 election, political and policy aspects of the 9/11 attacks and the invasion of Iraq. Tough, smart, hard working and well connected, she had rightfully earned a place near the top of one of the country's most established and respected media outlets. When we met after she had spent nearly thirty years at the *Post*, Maralee more or less epitomized what had come to be known as the "mainstream media."

Unsurprisingly, Maralee regarded the new media in general—and in particular the emerging social media I was examining—as some form of journalistic joke, and certainly not on a par with the lofty likes of the *Washington Post*. She also made it clear—from an initial meeting when she sniffed at my mention of social networks and asked if I was referring to "those places people go to get dates?" to our farewell dinner months later, when she hesitantly wished me "good luck in the future doing that sort of journalism that you do," that there was little room in her world for my outré inquiries and shadowy suggestions that social media might help fix journalism's crisis of credibility.

Yet for all her defensiveness—in fairness, it was not an uncommon response at the time among veteran journalists suffering through the precipitous decline of their profession—Maralee was at least open enough to concede that the entire media world was in the midst of rapid and revolutionary change. One day she knocked loudly on my office door, and when I opened it, half-shoved a young man in to meet me.

"This is Jose Antonio Vargas," she announced. "*He* gets what you are doing!"

Vargas, a young reporter from the *Post*, had been part of a team that earned a Pulitzer Prize for covering the 2007 shootings at Virginia Tech. In one article, "Students Make Connections at a Time of Total Disconnect," he reported on the role of technology in students' experiences during the shootings, and described how graduate student Jamal

Albarghouti, running towards the gun shots when he heard them, took out his cell phone to take a video that would later air on CNN.com. "This is what this YouTube-Facebook-instant messaging generation does," Vargas wrote. "Witness. Record. Share."

Maralee was right—Vargas *did* get it, so much so that within minutes of our meeting, he began to denounce his superiors at the *Post* for being way behind the digital information curve. "I don't understand why they won't just fire a bunch of the editors and replace them with people who know how to display digital information in a more graphic and friendly form!" he complained. When I suggested that voicing such sentiments too loudly or too often wouldn't earn him many friends in the Post newsroom, Vargas didn't seem overly concerned. A few months later—already a mainstream star at the tender age of 27—he left to take a job at the Huffington Post, one of the new media operations that was even then beginning to surpass the *Washington Post* in online page views, if not influence and prestige as well.

I saw Maralee Schwartz again three years later, during the Shorenstein Center's twenty-fifth anniversary celebration in October 2011. As I arrived at the festivities, I heard my name called from across the room and saw her gesturing for me to come over. She asked if I remembered what her original reaction had been to my research thesis.

"Yes, I do," I replied, nonplussed. "You were . . . somewhat skeptical, as I recall."

"Skeptical? I thought you were crazy!" Maralee said in

her usual straightforward, no-holds-barred manner. "But now I see you were completely right!"

It had only taken three years—an eternity in Internet time, where everything moves at the speed of light, but not really so long in the overall scheme of things—for Maralee to move from stalwart opponent to zealous enlistee and foot soldier in the digital information revolution. So yes, there is reason for hope and optimism as we face the many challenges of that ongoing struggle. . . . If prominent individuals like Maralee Schwartz, refugees from lifelong careers in the legacy media, can admit to the possibilities of change, so can our leading institutions, as Ethan Zuckerman and many others suggest. Both social and technological solutions to the credibility crisis and the revenue rout afflicting our media already exist and are constantly being refined and improved—and with them, our own sense of whom and what to believe is getting better all the time as well.

In any event the old legacy brands can't and won't provide much of an answer. There's simply no going back, because media doesn't just come from the media anymore. We are all the media now, and are unwilling to go through old media brands to access content or reach an audience, as we live through the process of what blogger Jeff Jarvis has called "the complete and utter disaggregation and disintermediation of media, turning everything about it upside-down." In this new, more social media environment, "content starts with the consumer instead of the creator;

authority is established by the public instead of the brand; the audience is the distributor," as Jarvis notes.

So there's no longer any need to imagine a media world where you create, aggregate and share freely and find credible, relevant news and information by using recommendations from peers you trust—because that world is already here. The feed really is your friends and followers!

Index

Acknowledgments

Many individuals helped me in the creation of this book, and I would like to acknowledge and thank them all. Their comments, insights and gracious assistance were instrumental in shaping it. They include:

Interviewees: Biz Stone, co-founder, Twitter; Robin Sloan, media strategist, Twitter; Randi Zuckerberg, former executive, Facebook; Reid Hoffman, founder, LinkedIn; Steve Grove, Director of *News & Politics*, YouTube; Andrew Heyward, ex-President, CBS News; Mark Lukasiewicz, NBC News; Paul Slavin, ABC News; Josh Tyrangiel, editor, Bloomberg BusinessWeek; Richard Gingras, ex-CEO, *Salon* magazine; Jose Antonio Vargas, former reporter, Huffington Post, *Washington Post*; Hemanshu Nigam, former executive, NewsCorp and Microsoft; Maralee Schwarz, ex-editor, *Washington Post*, Harvard University; Ethan Zuckerman, Massachusetts Institute of Technology, Harvard University; Robert Putnam, Harvard University; Tom Sander, Harvard University; Howard Gardner, Harvard University; BJ Fogg, Stanford University; Dean Eckles, Standford University; Paul Resnick, University of Michigan; Natalie Jomini Stroud, University of Texas at Austin; Danah Boyd, Microsoft, Harvard's University; Judith Donath, Massachusetts Institute of Technology, Harvard University; Eszter Hargittai, Northwestern University; Miriam Metzger, University of California, Santa Barbara; R. Kelly Garrett, The Ohio State University; Nicole Ellison, Michigan State University; Cliff Lampe, Michigan State University; the late Peresephone Miel, Harvard University; the late Rob Stuart, digital strategist; Marc Rotenberg, Electronic Privacy Information Center; Eli Pariser, author, *Filter Bubbles*; Jonathan Peizer, digital consultant; Mindy Finn, Republican campaign strategist; Nicco Mele, Harvard University; Jim Brayton, Internet advisor, Howard Dean & Barack Obama campaigns; Fabrice Florin, founder, NewsTrust.net; Richard Sambrook, Chief Content Officer, Edelman Public Relations; Robin Miller, editor, Slashdot.

Others whose work, insights, and/or assistance are also appreciated include: Ken Auletta, Craig Newmark, Dan Gillmor, Jeff Jarvis, Saul Hansell, Brian Reich, Peter Hart and Jay Campbell, Andy Carvin, Mark McKinnon, Dan Kennedy, Michael Wolff, J. Max Robins, and Andrew Lih.

Thanks also to my many friends and colleagues at the Joan Shorenstein Center on the Press, Politics and Public Policy, including Alex S. Jones, Nancy Palmer, Thomas Patterson, and Lorie Conway, Edith Holway, Richard Parker, Janell Sims, Leighton Klein, *et al*; to Paul Sagan for sponsoring my fellowship; to my co-fellows Sandra Nyaira, Eric Pooley, and Ed Shumacher-Matos for their fellowship, and to my fantastic research assistant at Harvard, Carrie Sheffield.

Special thanks are owed to Philip Balboni, Lisa Schmid Alford and Joel Alford, Arachu Castro, Patrice O'Neill, and particularly to Paul Solman and Jan Freeman for their continuing friendship, support—and free lodging!

Thank you also to Alison Humes, Angela Heimburger, Rose Lichter-Marck, Aaron Cutler, Sarah Lazin, Peter Richardson, and David Clark for editorial assistance—and of course to the great team at City Lights, including Greg Ruggiero, Stacey Lewis, and Alex.

Also thank you to my colleagues at Globalvision: Danny Schechter, Glenn Beatty, Pat Horstman, David DeGraw and David Neumann for their help—and finally to my many friends and followers on Facebook, Twitter, LinkedIn, and elsewhere in cyberspace!

About the Author

Rory O'Connor is a writer, filmmaker, and journalist whose work centers around media and politics. Author of *Shock Jocks: Hate Speech & Talk Radio*, and co-author of *Nukespeak: The Selling of Nuclear Power from the Manhattan Project to Fukushima*, his broadcast, film, and print career has been recognized with a George Orwell Award, a George Polk Award, a Writer's Guild Award, and two Emmys, among other honors.

Co-founder and president of the international media firm Globalvision, Inc, and Board Chair of The Global Center, an affiliated non-profit foundation, O'Connor has been a key figure in the production of dozens of documentary films and has also been an executive in charge of three weekly television series. Globalvision films and television programming have aired on leading broadcast and cable networks in more than one hundred countries—from ABC, NBC, CBS, PBS, and FOX domestically to the BBC, RAI, NHK, National Geographic, and many others internationally.

A longtime blogger for sites such as the Huffington Post, AlterNet, and others, including his own Media Is A Plural, he has also appeared as an on-air commentator and "vlogger" on international broadcast systems such as Al Jazeera and the CBC. His website is www.roryoconnor.org.